REVISE EDEXCEL GCSE (9–1)
Geography A

REVISION WORKBOOK

Series Consultant: Harry Smith

Author: Alison Barraclough

A note from the publisher

In order to ensure that this resource offers high-quality support for the associated Pearson qualification, it has been through a review process by the awarding body. This process confirms that this resource fully covers the teaching and learning content of the specification or part of a specification at which it is aimed. It also confirms that it demonstrates an appropriate balance between the development of subject skills, knowledge and understanding, in addition to preparation for assessment.

Endorsement does not cover any guidance on assessment activities or processes (e.g. practice questions or advice on how to answer assessment questions), included in the resource nor does it prescribe any particular approach to the teaching or delivery of a related course.

While the publishers have made every attempt to ensure that advice on the qualification and its assessment

is accurate, the official specification and associated assessment guidance materials are the only authoritative source of information and should always be referred to for definitive guidance.

Pearson examiners have not contributed to any sections in this resource relevant to examination papers for which they have responsibility.

Examiners will not use endorsed resources as a source of material for any assessment set by Pearson.

Endorsement of a resource does not mean that the resource is required to achieve this Pearson qualification, nor does it mean that it is the only suitable material available to support the qualification, and any resource lists produced by the awarding body shall include this and other appropriate resources.

For the full range of Pearson revision titles across KS2, KS3, GCSE, Functional Skills, AS/A Level and BTEC www.pearsonschools.co.uk/revise

Pearson

CONTENTS

COMPONENT 1: THE PHYSICAL ENVIRONMENT

Changing UK landscapes
1 Main UK rock types
2 Upland and lowland landscapes
3 Physical processes
4 Human activity

Coastal landscapes
5 Physical processes 1
6 Physical processes 2
7 Influence of geology
8 UK weather and climate
9 Erosional landforms
10 Depositional landforms
11 Human activity
12 Coastal management
13 Holderness coast

River landscapes
14 Physical processes 1
15 Physical processes 2
16 River valley changes
17 Weather and climate challenges
18 Upper course landscape
19 Lower course landscape 1
20 Lower course landscape 2
21 Human impact
22 Causes and effects of flooding
23 River management
24 River Dee, Wales

Glaciated upland landscapes
25 Glacial processes
26 Erosion landforms 1
27 Erosion landforms 2
28 Transport and deposition landforms
29 Human activity
30 Glacial development

Weather and climate
31 Global atmospheric circulation
32 Natural climate change
33 Human activity
34 The UK's climate
35 Tropical storms
36 Tropical cyclone hazards
37 Hurricane Sandy
38 Typhoon Haiyan
39 Drought causes and locations
40 California, USA
41 Ethiopia

Ecosystems
42 The world's ecosystems
43 Importance of the biosphere
44 The UK's main ecosystems
45 Tropical rainforest features
46 TRF biodiversity and adaptations
47 TRF goods and services
48 Deforestation in tropical rainforests
49 Tropical rainforest management
50 Deciduous woodlands features
51 Deciduous woodlands adaptations
52 Deciduous woodlands goods and services
53 Deforestation in deciduous woodlands
54 Deciduous woodlands management

Extended writing questions
55 Paper 1

COMPONENT 2: THE HUMAN ENVIRONMENT

Changing cities
56 An urban world
57 UK urbanisation differences
58 Context and structure
59 A changing UK city
60 Globalisation and economic change
61 City inequalities
62 Retailing changes
63 City living
64 Context and structure
65 A rapidly growing city
66 Increasing inequalities
67 Solving city problems

Global development
68 Defining development
69 Measuring development
70 Patterns of development
71 Uneven development
72 International strategies
73 Top-down vs bottom-up
74 Location and context
75 Uneven development and change
76 Trade, aid and investment
77 Changing population
78 Geopolitics and technology
79 Impact of rapid development

Resource management
80 The world's natural resources
81 Variety and distribution
82 Global usage and consumption

Energy
83 Production and development
84 UK and global energy mix
85 Impacts of non-renewable energy resources
86 Impacts of renewable energy resources
87 Meeting energy demands
88 China and Germany

Water
89 Global distribution of water
90 Changing water use
91 Water consumption differences
92 Water supply problems: UK
93 Water supply problems: emerging or developing countries
94 Attitudes and technology
95 Managing water
96 UK and China

Extended writing questions
97 Paper 2

COMPONENT 3: GEOGRAPHICAL INVESTIGATIONS

Fieldwork: coasts
98 Formulating enquiry questions
99 Methods and secondary data
100 Working with data

Fieldwork: rivers
101 Formulating enquiry questions
102 Methods and secondary data
103 Working with data

Fieldwork: urban
104 Formulating enquiry questions
105 Methods and secondary data
106 Working with data

Fieldwork: rural
107 Formulating enquiry questions
108 Methods and secondary data
109 Working with data

UK challenges
110 Consumption and environmental challenges
111 Population and economic challenges
112 Landscape challenges
113 Climate change challenges

Extended writing questions
114 Paper 3 (i)
115 Paper 3 (ii)

GEOGRAPHICAL, MATHEMATICS AND STATISTICS SKILLS
116 Atlas and map skills
117 Types of map and scale
118 Using and interpreting images
119 Sketch maps and annotations
120 Physical and human patterns
121 Land use and settlement shapes
122 Human activity and OS maps
123 Map symbols and direction
124 Grid references and distances
125 Cross-sections and relief
126 Graphical skills 1
127 Graphical skills 2
128 Graphical skills 3
129 Numerical and statistical skills 1
130 Numerical and statistical skills 2
131 Resources for Paper 3 (ii)
132 Answers

A small bit of small print
Edexcel publishes Sample Assessment Material and the Specification on its website. This is the official content and this book should be used in conjunction with it. The questions have been written to help you practise every topic in the book. Remember: the real exam questions may not look like this.

Main UK rock types

1 Study **Figure 1**.

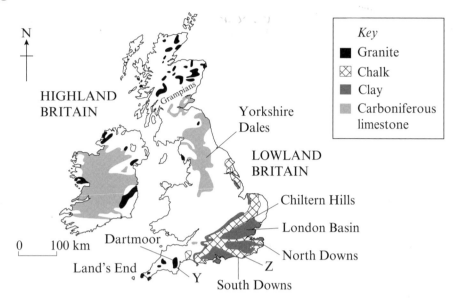

Figure 1 Selected rock types and upland areas in the UK

(a) Identify the rock name and rock group found at **Y** and **Z**, and write them in the table below.

Letter	Rock name	Rock group
Y	Granite	Igneous
Z	Chalk	Sedimentary

                                                                    (4 marks)

(b) State **one** characteristic of an igneous rock.

.... They   are   impermable .... **(1 mark)**

(c) Identify **one** characteristic of metamorphic rocks.

☐ **A** Often contain fossils

☐ **B** Formed from magma

☑ **C** Formed by heat and pressure

☐ **D** Made up of calcium carbonate   **(1 mark)**

(d) Identify the sedimentary rock that is made up of grains of sand cemented together.

.... sandstone .... **(1 mark)**

⟩ **Guided** ⟩   2 Explain **one** characteristic of granite landscapes.

> Remember that **explain** means you have to give reasons.

Granite is impermeable. This means ..... that   water   cannot go

.... through. .... **(2 marks)**

Upland and lowland landscapes

1   Identify the imaginary line that divides the UK into upland and lowland areas.

...........Tees- Exe.. **(1 mark)**

2   Study **Figure 1**.

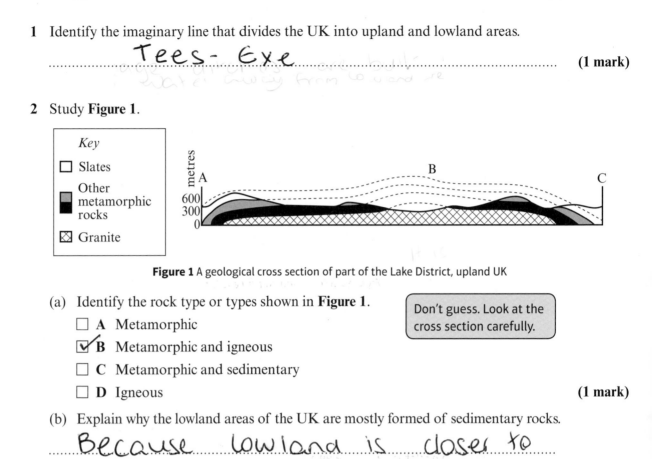

Figure 1 A geological cross section of part of the Lake District, upland UK

Key
- ☐ Slates
- ▨ Other metamorphic rocks
- ⊠ Granite

(a)   Identify the rock type or types shown in **Figure 1**.

☐ **A**   Metamorphic

☑ **B**   Metamorphic and igneous

☐ **C**   Metamorphic and sedimentary

☐ **D**   Igneous   **(1 mark)**

> Don't guess. Look at the cross section carefully.

(b)   Explain why the lowland areas of the UK are mostly formed of sedimentary rocks.

....Because lowland is closer to sea level ~~and~~, sand and beaches **(2 marks)**

> **Guided**

3   Study **Figure 2**.

Figure 2 Part of the Giant's Causeway, Northern Ireland

Suggest how the polygon rock shapes shown in **Figure 2** are formed.

Diverging plate boundaries about 60 million years ago meant that the Atlantic

Ocean began to form. Rising lava ....

.... **(2 marks)**

Physical processes

1 Study **Figure 1**.

Scree slope

Stickle Tarn in a corrie

Figure 1 Stickle Tarn, in the Lake District

(a) Identify the physical process by which the scree slope in **Figure 1** has been formed.

☑ **A** Erosion

☐ **B** Deposition

☐ **C** Precipitation

☐ **D** Weathering   **(1 mark)**

▷ Guided ▷   (b) Explain how the corrie containing Stickle Tarn shown in **Figure 1** was formed.

The corrie containing Stickle Tarn was formed by glacial ...... *erosion, where*
a   glacier had formed   during the Ice Age   **(2 marks)**
and   carved   our a   corrie.

2 Study **Figure 2**.

Figure 2 The confluence (joining point) of the rivers Wye and Lugg

Explain how the lowland landscape shown in **Figure 2** was formed by the interaction of physical processes.

> The question asks for **processes**, so it's important to include **more than one** physical process.

This landscape was affected by river erosion.
This means that as the river meandered it
eroded a valley into the low hills. It was also
affeated by river deposition. This means that heavy   **(4 marks)**
rain flooded the river and spread out onto
the valley,   creating a flood plain

3

Human activity

1 (a) State **one** way that agriculture has affected the UK landscape.

Drainage ditches are built to remove **(1 mark)**
excess water away from lowland areas to farm

(b) Explain **one** change to the landscape caused by forestry.

....

.... **(2 marks)**

Guided 2 Explain why farming has produced different distinctive landscapes in upland and lowland UK.

There are different types of farming in the UK's upland and lowland landscapes.

Steep slopes and poor soils in upland areas mean that ....It is better

suited to farm and breed...

.... **(4 marks)**

3 Study **Figure 1**.

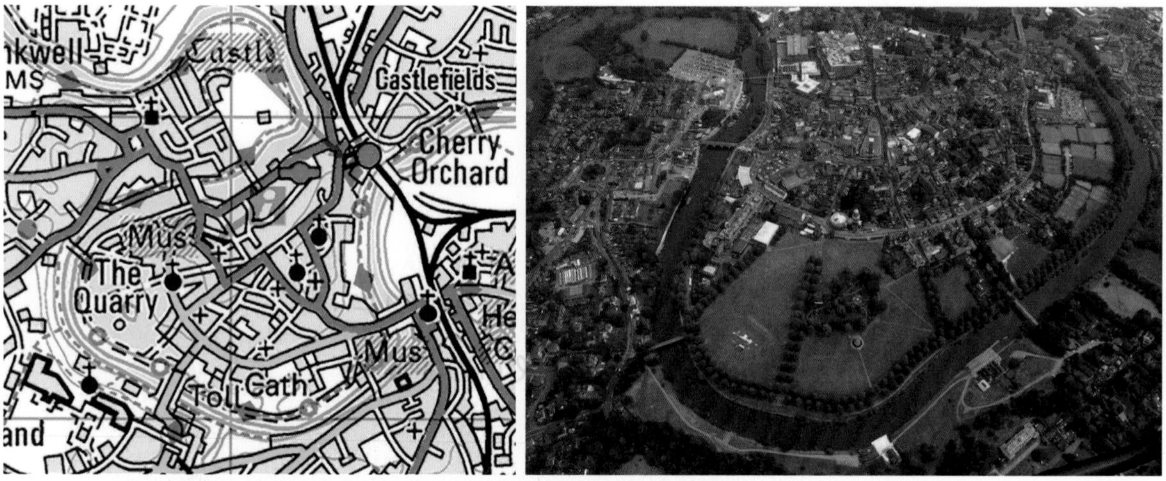

Figure 1 An Ordnance Survey extract (1:50 000) and aerial photograph showing part of Shrewsbury, Shropshire

Ordnance Survey Maps, © Crown copyright 2017, OS 100030901 and supplied by courtesy of Maps International.

(a) Identify **one** reason why the area shown in **Figure 1** might **not** be a suitable site for a settlement.

☐ **A** Risk of flooding   ☐ **C** Easy to build bridges

☐ **B** Flat land   ☐ **D** The river meanders   **(1 mark)**

(b) Suggest **one** way settlements such as Shrewsbury create a distinctive landscape.

> Look carefully at the resource for ideas.

....

.... **(2 marks)**

4 Explain why upland landscapes have been altered less by human settlement than have lowland landscapes.

....

....

.... **(4 marks)**

Physical processes 1

Only revise pages 5–13 if you studied Coastal landscapes.

1   Identify the process by which rocks are broken down in situ.

.... **(1 mark)**

> **Guided**

2   Study **Figure 1**.

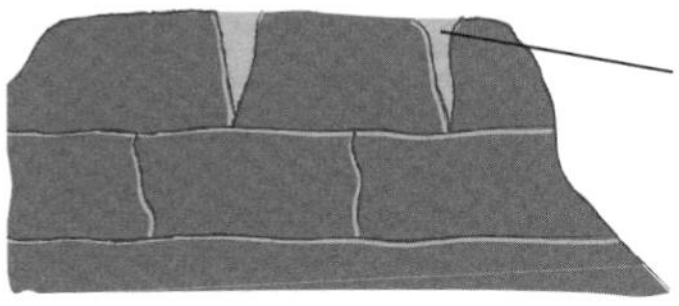

Water freezes and the crack is widened.

Figure 1 Some information about how rocks are broken down

Explain the process shown in **Figure 1**.

.... The water freezes and the crack is widened. ....

.... **(2 marks)**

3   Which **one** of the following is the term for the downhill movement of material due to gravity?

☐ **A** Mass wasting   ☐ **C** Downhill movement

☐ **B** Gravity flow   ☐ **D** Mass movement   **(1 mark)**

4   Study **Figure 2**.

Figure 2 Waves eroding part of the UK coastline

Explain **two** ways that a coastline can be eroded by wave action.

Only explain processes that **cause erosion of the coast**, not the processes that break up sediment that has already been eroded.

....

....

....

....

.... **(4 marks)**

Physical processes 2

1 Study **Figure 1**.

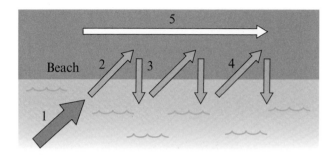

Figure 1 The movement of sediment along a beach

(a) State the term for the movement of sediment along a beach.

.... **(1 mark)**

(b) Identify the process shown by **arrow 3** in **Figure 1.**

.... **(1 mark)**

> **Guided** 2 Explain how waves transport material.

The largest material, such as boulders, is rolled ....

.... **(2 marks)**

3 Study **Figure 2**.

Figure 2 A beach in the UK

(a) Identify the type of wave labelled **A** in **Figure 2**.
    Look carefully at the whole photo. The amount of beach material that has built
    up provides a clue.

.... **(1 mark)**

(b) Explain the process of deposition of material (load) transported by waves.

....

.... **(2 marks)**

Influence of geology

1   Study **Figure 1**.

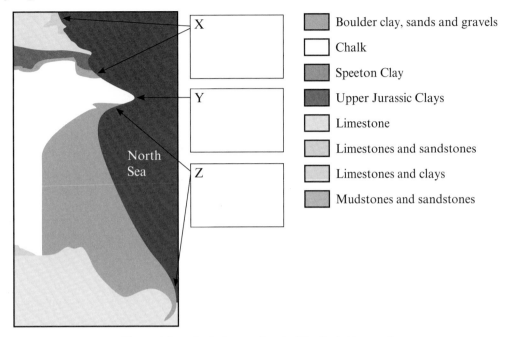

■	Boulder clay, sands and gravels
☐	Chalk
■	Speeton Clay
■	Upper Jurassic Clays
■	Limestone
■	Limestones and sandstones
■	Limestones and clays
■	Mudstones and sandstones

Figure 1 A geological map of part of the Yorkshire coast

(a)   Label **boxes X, Y and Z** on **Figure 1** to indicate which is: (i) a section of
concordant coast, (ii) a section of discordant coast and (iii) a headland.   **(3 marks)**

(b)   Suggest how the landforms vary between a coastline made up of hard rocks and
a coastline made up of soft rocks.

....

....   **(2 marks)**

>**Guided**   2   State **two** differences between destructive and constructive waves.

Destructive waves are ....

However, constructive waves are low-energy waves that ....

....   **(2 marks)**

3   Study **Figure 2**.

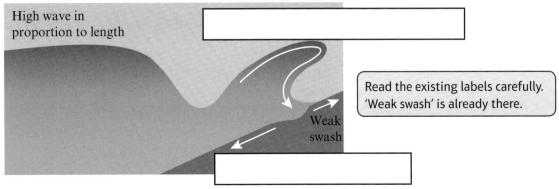

High wave in
proportion to length

Read the existing labels carefully.
'Weak swash' is already there.

Weak
swash

Figure 2 A destructive wave

Add labels to the blank boxes in **Figure 2** to describe the characteristics of a
destructive wave.   **(2 marks)**

UK weather and climate

1 (a) Prevailing winds affect the UK. Identify the direction from which these prevailing winds come.

☐ **A** North-east   ☐ **C** North-west

☐ **B** South-west   ☐ **D** South-east   **(1 mark)**

(b) Identify the type of weather prevailing winds bring to the UK.

.... **(1 mark)**

2 Identify the UK's climatic type.

.... **(1 mark)**

3 State **one** type of weathering that affects rocks along the coastline.

.... **(1 mark)**

〉**Guided**〉 4 Study **Figure 1**.

Figure 1 Coastal erosion in Yorkshire

(a) Define the term **coastal erosion**.

> Make sure you know the difference between weathering and erosion.

.... **(1 mark)**

(b) Suggest **two** impacts of coastal erosion on the UK coastline.

Coastal erosion can cause coastal recession, which is when the coastline moves

.... **(2 marks)**

(c) Explain how strong winds and storms can increase the rate of coastal erosion.

....

....

....

.... **(4 marks)**

Erosional landforms

1 Study **Figure 1**.

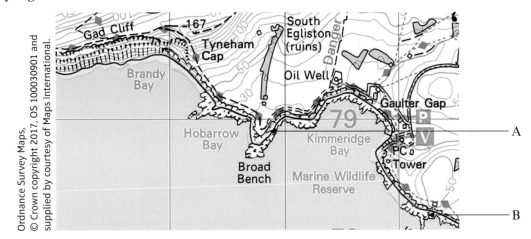

Ordnance Survey Maps,
© Crown copyright 2017, OS 10003090I and
supplied by courtesy of Maps International.

Figure 1 1:50 000 Ordnance Survey map of Kimmeridge Bay, Dorset

Identify the landforms found at **A** and **B** in **Figure 1**.

A ....   B ....   **(2 marks)**

Guided

2 Study **Figure 2**.

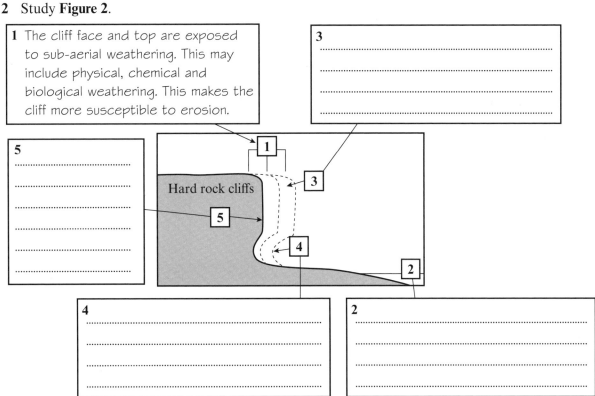

1 The cliff face and top are exposed to sub-aerial weathering. This may include physical, chemical and biological weathering. This makes the cliff more susceptible to erosion.

3 ....

5 ....

Hard rock cliffs

4 ....

2 ....

Figure 2 The formation of a wave cut platform

Explain the formation of a wave cut platform by completing **boxes 2–5** on **Figure 2**.   **(4 marks)**

3 Describe how an arch is formed in a coastal headland.

> There are two marks, so make sure you describe at least two stages.

....

....   **(2 marks)**

9

Depositional landforms

1   Study **Figure 1**.

Figure 1 1:50 000 Ordnance Survey map of Slapton Ley, Devon

(a)   Identify the landform found at **A** on **Figure 1**.

....

> Make sure you can recognise coastal landforms on Ordnance Survey maps.

(1 mark)

(b)   State **two** features of this landform.

....

....

(2 marks)

2   Explain why beaches can be formed of different sediment types.

....

....

(2 marks)

> **Guided**

3   Study **Figure 2**.

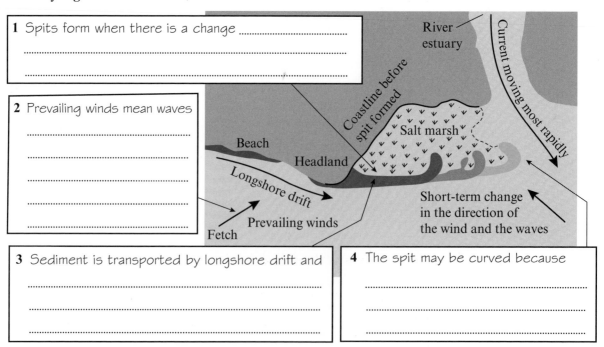

1 Spits form when there is a change ....

....

....

2 Prevailing winds mean waves

....

....

....

....

....

3 Sediment is transported by longshore drift and

....

....

....

4 The spit may be curved because

....

....

....

River estuary

Current moving most rapidly

Coastline before spit formed

Salt marsh

Beach

Headland

Longshore drift

Prevailing winds

Fetch

Short-term change in the direction of the wind and the waves

Figure 2 The formation of a spit

Explain the formation of a spit by completing **boxes 1–4** on **Figure 2**.   **(4 marks)**

Human activity

1   Study **Figure 1**.

Figure 1 A section of the Yorkshire coast

> Guided

(a) Name the coastal process shown in **Figure 1**.

Coastal ....   **(1 mark)**

(b) Explain **one** way the process shown in **Figure 1** might affect local people.

....

....   **(2 marks)**

(c) Suggest how the cliff-top settlement might affect the rate of erosion.

....

....

....

....   **(4 marks)**

2   Explain **one** way in which industry can change the coastal environment.

> Changes may be positive as well as negative.

....

....   **(2 marks)**

3   Explain how coastal recession and flooding can affect the environment.

....

....

....

....   **(4 marks)**

Coastal management

1   Groynes are an example of a hard engineering technique that helps to prevent coastal erosion. State **one** advantage and **one** disadvantage of using groynes.

Advantage ....

Disadvantage .... **(2 marks)**

2   Study **Figure 1**.

Figure 1 Coastal protection at Scarborough, Yorkshire

(a)   Identify the coastal protection technique labelled **A** in **Figure 1**.

☐ **A**   A groyne   ☐ **C**   An offshore reef

☐ **B**   A sea wall   ☐ **D**   Rip rap   **(1 mark)**

Guided (b)   Explain how this coastal protection technique helps to prevent coastal erosion.

This technique reduces wave energy because ....

.... **(2 marks)**

(c)   Identify the coastal protection technique labelled **B** in **Figure 1**.

☐ **A**   An embankment   ☐ **C**   An offshore reef

☐ **B**   A sea wall   ☐ **D**   Rip rap   **(1 mark)**

(d)   Explain how this coastal protection technique helps to prevent coastal erosion.

....

.... **(2 marks)**

3   Explain **one** advantage and **one** disadvantage of using soft engineering as a form of coastal management.

> Make sure you refer to specific types of soft engineering in your answer.

....

....

....

.... **(4 marks)**

🌐 Located example **Holderness coast**

> **Guided**

1 (a) Study **Figure 1**.

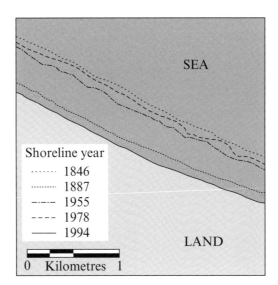

Figure 1 A diagram showing the rate of coastal recession along a section of the Holderness coast, Yorkshire

Describe the rate of coastal recession shown in **Figure 1**.
Use data in your answer.

Between 1846 and 1994, the coastline receded by more than half a kilometre.

.... **(2 marks)**

(b) Study **Figure 2**.

Section of the Holderness coast	Annual rate of coastal recession (metres)
North	1.72
Centre	1.53
South	2.08

Figure 2 Coastal erosion rates, the Holderness coast

Using the information in **Figure 2**, calculate the
mean rate of coastal recession for the
Holderness coast.

> To calculate the mean, add up
> the figures and divide by 3.

.... **(1 mark)**

2 For a named distinctive coastal landscape, examine influential factors that are causing
it to change.

Named coastal landscape ....

.... **(8 marks)**

....

....

> Continue your answer on your own paper. You should aim to write approximately one side of A4.

Physical processes 1

> Only revise pages 14–24 if you studied River landscapes.

1 Define the term **weathering**.

.... **(1 mark)**

〉 **Guided** 〉 2 Mechanical weathering is one type of weathering. Name the other **two** types of weathering.

(a) Biological   (b) ....   **(1 mark)**

3 Study **Figure 1**.

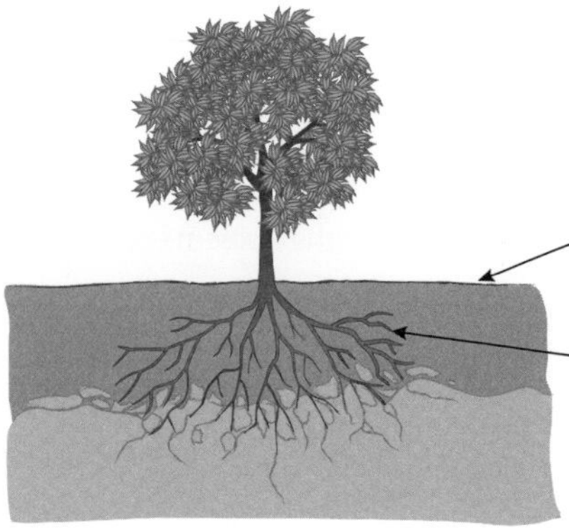

1 Acids ....
....
....

2 Roots ....
....
....

Figure 1 Biological weathering

Explain the process of chemical weathering by completing **boxes 1** and **2** on **Figure 1**.   **(4 marks)**

4 Study **Figure 2**.

— Mass
movement

Figure 2 The upper course of a river in Glencoe, Scotland

Explain the process of mass movement
that occurs in the upper courses of rivers.

> Figure 2 is only there to help you. You are not
> expected to know about any specific rivers.

....

.... **(2 marks)**

Physical processes 2

1 Study **Figure 1**.

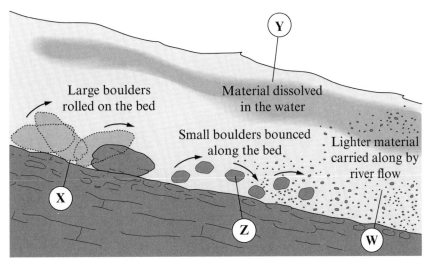

Figure 1 The different ways a river transports its load

(a) Name the processes shown at **W, X, Y** and **Z** on **Figure 1**.

W ....

X ....

Y ....

Z .... **(4 marks)**

(b) Which **one** of the processes shown on **Figure 1** does **not** involve erosion?

.... **(1 mark)**

2 Define the term **transport**.

> When you are asked to define a term you need to say exactly what the term means.

.... **(1 mark)**

3 Identify the process that takes place when a river loses energy.

☐ **A** Erosion

☐ **B** Transport

☐ **C** Weathering

☐ **D** Deposition **(1 mark)**

Guided

4 Explain the difference between abrasion and attrition.

Abrasion and attrition both involve erosion. However, abrasion takes place when

the river's load rubs along the riverbed and banks, ...

....

.... **(4 marks)**

River valley changes

1 Study **Figure 1**.

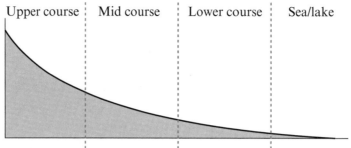

Figure 1 The long profile of a river

(a) Define the term **long profile of a river**.

.... **(1 mark)**

(b) Identify the section of a river's long profile in which it has the steepest gradient.

☐ **A** Upper course   ☐ **C** Lower course

☐ **B** Mid course   ☐ **D** Sea/lake   **(1 mark)**

(c) Identify the section of a river's long profile in which it has the highest velocity.

☐ **A** Upper course   ☐ **C** Lower course

☐ **B** Mid course   ☐ **D** Sea/lake   **(1 mark)**

2 Explain how a river's channel shape changes downstream.

> Use your knowledge of the long profile of a river to help you.

....

.... **(2 marks)**

> **Guided**

3 Explain **two** ways that river landscapes contrast between the upper and lower sections of the long profile.

The landscape of the upper section of a river is formed by erosion ....

....

However, the lower section of a river valley ....

.... **(2 marks)**

4 For a named UK river you have studied, explain how the local geology affects its long profile.

Named UK river ....

....

....

....

.... **(4 marks)**

Weather and climate challenges

1　Study **Figure 1**.

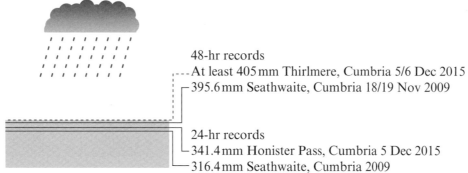

48-hr records
At least 405 mm Thirlmere, Cumbria 5/6 Dec 2015
395.6 mm Seathwaite, Cumbria 18/19 Nov 2009

24-hr records
341.4 mm Honister Pass, Cumbria 5 Dec 2015
316.4 mm Seathwaite, Cumbria 2009

Figure 1 Information about short-term periods of intense rainfall

(a)　Which **one** of the following had the highest mean rainfall **during a 24-hour period**?

> To work out the mean, add the numbers together and divide the total by the amount of numbers.

　☐ **A**　Thirlmere, December 2015　　☐ **C**　Honister Pass, December 2015

　☐ **B**　Seathwaite, November 2009　　☐ **D**　Seathwaite, 2009　　**(1 mark)**

(b)　Suggest the impact that intense rainfall events such as those shown in **Figure 1** might have on local river discharge.

....

.... **(2 marks)**

Guided　2　Study **Figure 2**.

> **1** Erosion rates will be higher with greater discharge, so landforms like ....
> ....
> ....
> ....
> ....

> **2** Load transport will
> ....
> ....
> ....
> ....
> ....
> ....

Figure 2 An aerial photograph of the River Severn, Shropshire

Explain the impacts of a storm event on (a) river landforms and (b) load transport by completing **boxes 1** and **2** on **Figure 2**.　　**(2 marks)**

3　Explain how extreme weather events can increase the risk of river flooding.

....

....

.... **(3 marks)**

Had a go ☐ Nearly there ☐ Nailed it! ☐

Upper course landscape

1 Study **Figure 1.**

(a) Name the landforms labelled **A** in **Figure 1**.

.... **(1 mark)**

(b) Explain how the landforms labelled **A** formed.

....

....

....

....

.... **(4 marks)**

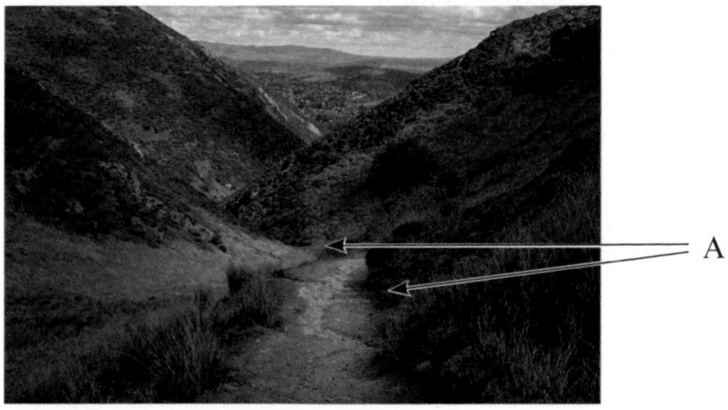

Figure 1 The upper course of a river

2 Study **Figure 2**.

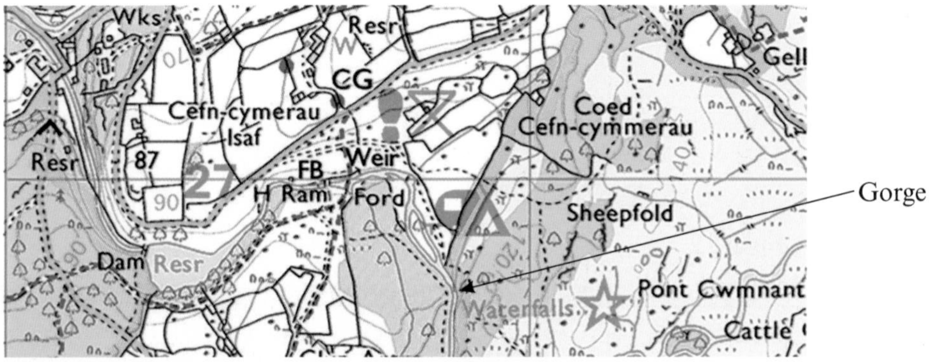

Figure 2 1:25 000 Ordnance Survey map of Afon (River) Cwmnantcol

(a) Explain the formation of a gorge.

Include geographical terms, such as abrasion.

....

....

....

.... **(4 marks)**

Guided

(b) Explain how erosional processes and geology work together to form the waterfall in **Figure 2**.

Waterfalls often occur where rocks of different hardness, and therefore resistance to erosion, occur together. The harder ....

....

.... **(4 marks)**

Lower course landscape 1

Guided 1 Study **Figure 1**.

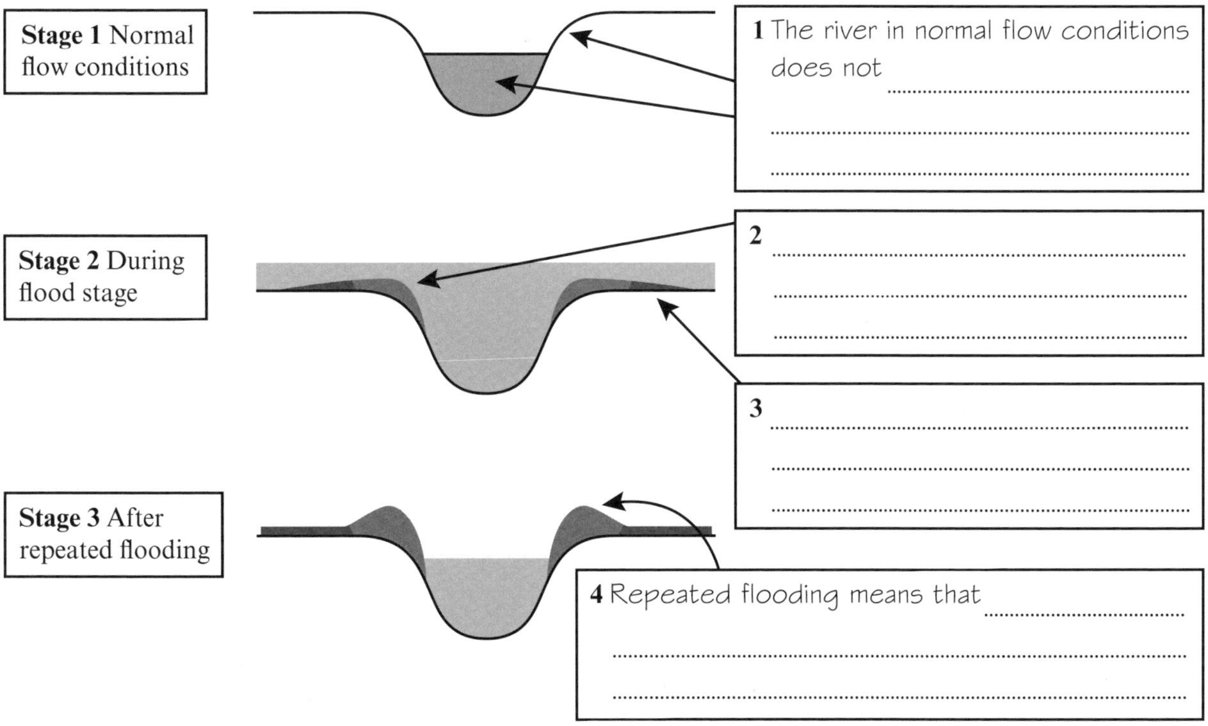

Stage 1 Normal flow conditions

1 The river in normal flow conditions does not ....

....

....

Stage 2 During flood stage

2 ....

....

....

....

3 ....

....

....

Stage 3 After repeated flooding

4 Repeated flooding means that ....

....

....

Figure 1 The formation of levées

Explain the formation of levées by completing **boxes 1–4** on **Figure 1**.   **(4 marks)**

2 Study **Figure 2**.

Figure 2 An aerial photograph of the River Cuckmere, Sussex

Identify the landforms at **A** and **B** in **Figure 2**. | Look carefully at the resource. The land at B is flat. |

A ....   B ....   **(2 marks)**

3 Explain the formation of landform **B** shown in **Figure 2**.

....

....

....

....   **(4 marks)**

Lower course landscape 2

1 (a) Identify the direction of erosion in the mid and lower sections of a river's course.

> Remember your map reading skills and look at the map carefully.

.... **(1 mark)**

(b) State **two** changes to river channels that occur in the mid course.

(i) .... (ii) .... **(2 marks)**

2 Study **Figure 1**.

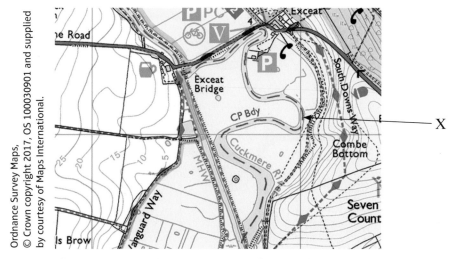

Ordnance Survey Maps,
© Crown copyright 2017, OS 100030901 and supplied by courtesy of Maps International.

Figure 1 1:25 000 Ordnance Survey map of River Cuckmere, Sussex

Identify the landform labelled **X** in **Figure 1**.

☐ **A** Oxbow lake ☐ **B** Floodplain ☐ **C** Meander ☐ **D** Levée **(1 mark)**

Guided **3** Study **Figure 2**.

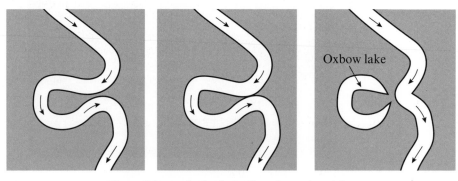

Figure 2 Stages in the formation of an oxbow lake

Examine how erosion and deposition work together in the formation of an oxbow lake.

When a river flows across its floodplain, it meanders. Deposition occurs

....

....

....

.... **(8 marks)**

> Continue your answer on your own paper. You should aim to write approximately one side of A4.

Human impact

1 Explain **one** way that agriculture can affect river processes.

> Make your answer specific to farming: for example, talk about ploughing.

....

.... **(2 marks)**

2 Identify **one** way that industry can affect river processes.

☐ **A** By abstracting water for irrigation

☐ **B** By abstracting water for domestic use

☐ **C** By abstracting water for manufacturing

☐ **D** By abstracting water for recreation **(1 mark)**

3 Which **one** of the following can happen as a result of abstracting water from a river?

☐ **A** Erosion decreases ☐ **C** Weathering increases

☐ **B** Erosion increases ☐ **D** Velocity increases **(1 mark)**

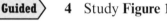 **4** Study **Figure 1**.

Figure 1 The River Kennet at Reading

Suggest how urbanisation, such as that shown in **Figure 1**, affects river processes.

Urbanisation has caused towns and cities to grow. Buildings are often constructed

on river floodplains, which means that ....

....

.... **(4 marks)**

5 Explain how human activities can affect the frequency of flood events.

....

....

....

.... **(4 marks)**

Causes and effects of flooding

1 Study **Figure 1**.

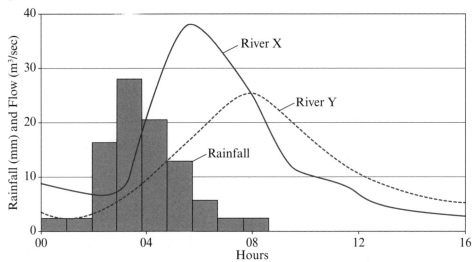

Figure 1 Storm hydrographs for two rivers

(a) Add labels for the following features on the storm hydrograph for River X in **Figure 1**.

> Remember that you might be asked to construct hydrographs in the exam.

   (i) Rising limb   (ii) Lag time   (iii) Peak rainfall   **(3 marks)**

> Guided

(b) Describe **two** differences in the storm hydrographs for **River X** and **River Y**. Use data in your answer.

The graph for River X has a much higher peak, rising to 38 m³/sec, but the peak

for River Y .... **(2 marks)**

(c) Define the term **lag time**.

.... **(1 mark)**

2 Study **Figure 2**.

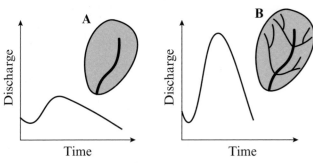

Figure 2 Storm hydrographs for two drainage basins

(a) Explain why the risk of flooding is greater in the drainage basin labelled **B** in **Figure 2**.

....

....

.... **(4 marks)**

(b) State **one** other physical factor that can increase the risk of flooding.

.... **(1 mark)**

River management

> **Guided**

1 (a) Define the term **hard engineering**.

Using man-made structures .... **(1 mark)**

(b) Explain **one** advantage OR **one** disadvantage of using hard engineering for flood defence.

> Make sure you know at least **one** advantage and **one** disadvantage for each flood defence technique.

....

....

.... **(2 marks)**

2 Study **Figure 1**.

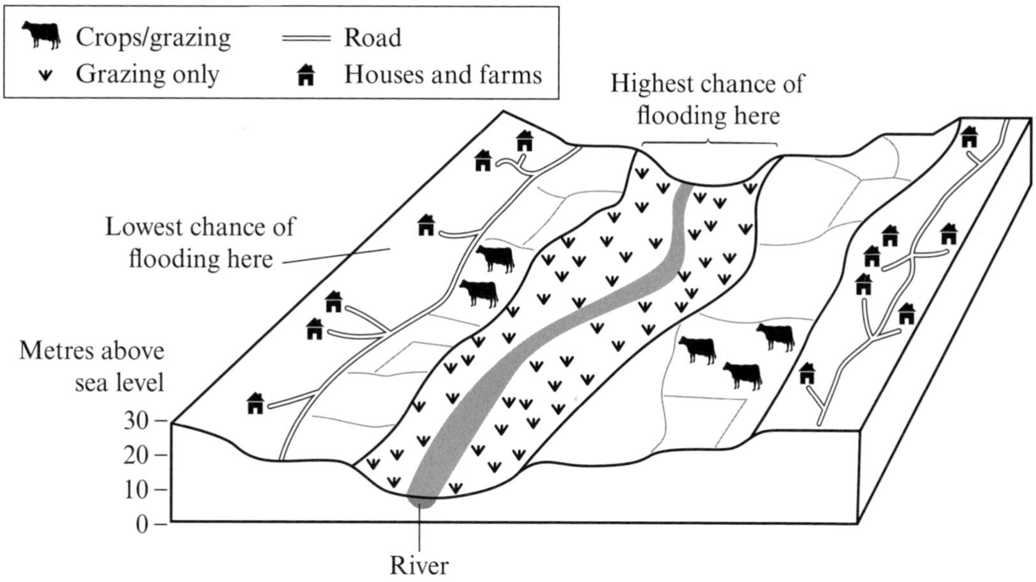

Key:
- 🐄 Crops/grazing   ═══ Road
- ⌄ Grazing only   🏠 Houses and farms

Highest chance of flooding here

Lowest chance of flooding here

Metres above sea level
30 –
20 –
10 –
0 –

River

Figure 1 A soft engineering flood protection method

(a) Name the soft engineering method shown in **Figure 1**.

.... **(1 mark)**

(b) Suggest how soft engineering techniques may lead to changes in river landscapes.

....

....

....

....

....

.... **(4 marks)**

Had a go ☐ Nearly there ☐ Nailed it! ☐

Located example River Dee, Wales

1 Identify the area which is the source of the River Dee.

☐ **A** An area of lowland glaciation

☐ **B** An area of sedimentary rocks

☐ **C** An area of upland glaciation

☐ **D** The Wirral estuary

(1 mark)

Guided

2 Study **Figure 1**.

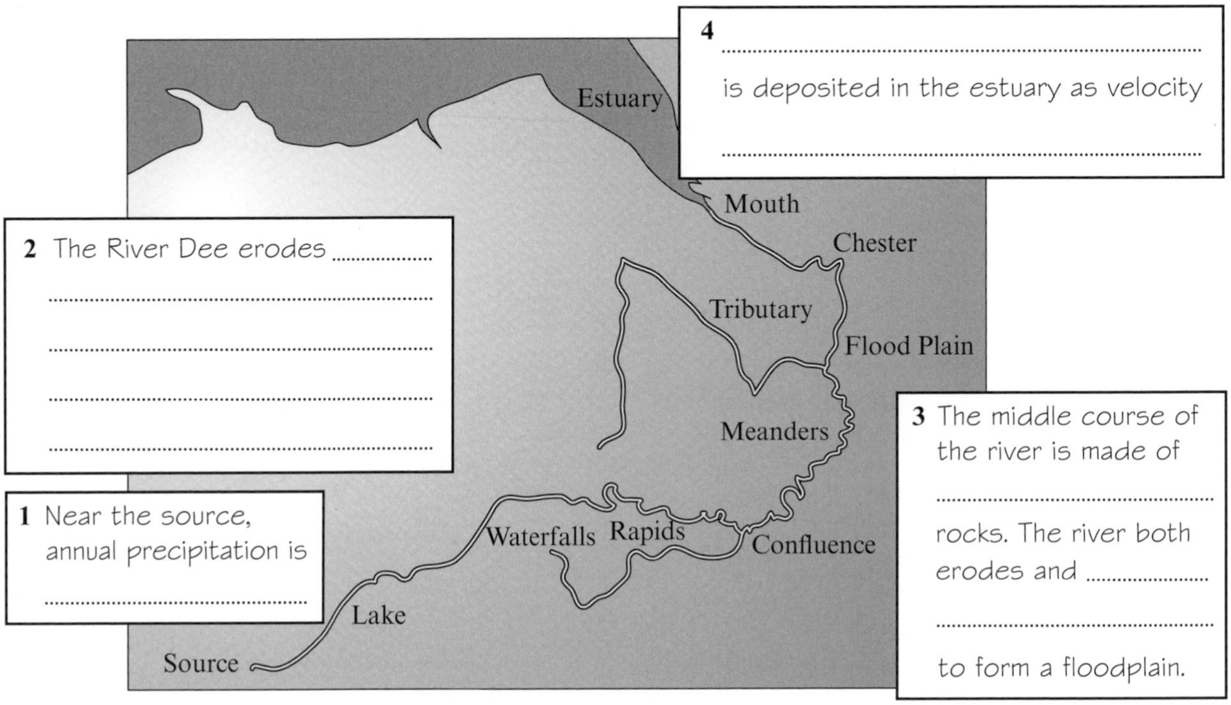
4 ....
is deposited in the estuary as velocity
....
Estuary
Mouth
Chester
Tributary
Flood Plain
Meanders
Confluence
Waterfalls
Rapids
Lake
Source
2 The River Dee erodes ....
1 Near the source, annual precipitation is ....
3 The middle course of the river is made of .... rocks. The river both erodes and .... to form a floodplain.

Figure 1 The course of the River Dee

Describe some of the characteristic features of the River Dee by completing **boxes 1–4** on **Figure 1**. **(2 marks)**

3 Examine how the interaction of physical and human processes worked together to form a named, distinctive river landscape.

> **Examine** means that you need to discuss individual factors such as geology, as well as the interaction of a range of factors.

Named distinctive river landscape ....

....

....

.... **(8 marks)**

> Continue your answer on your own paper. You should aim to write approximately one side of A4.

Glacial processes

Only revise pages 25–30 if you studied Glaciated upland landscapes.

1   Identify the type of weathering that occurs in glacial uplands.

.... **(1 mark)**

Guided   2   Study **Figure 1**.

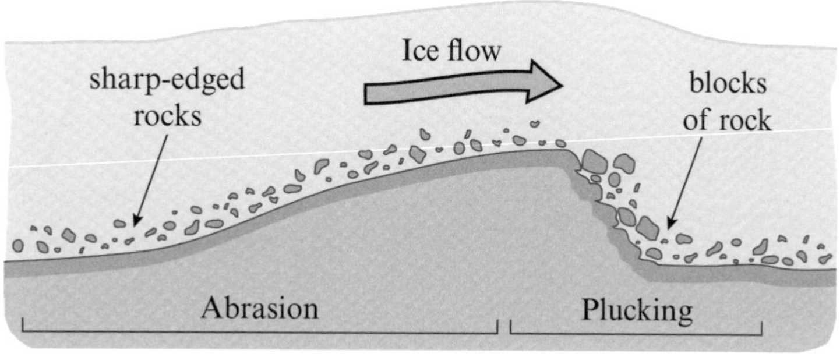

sharp-edged rocks

Ice flow

blocks of rock

Abrasion

Plucking

Figure 1 Glacial erosion

Explain the differences between the processes of abrasion and plucking.

Abrasion occurs when angular rocks become embedded in the base of the glacier.

As the glacier moves ....

However, plucking occurs when blocks of bedrock, loosened by freeze-thaw

weathering, freeze to the base .... **(4 marks)**

Guided   3   Study **Figure 2**.

1 Till deposits are ....
....
....

2 Fluvioglacial material is ....
....
....

Figure 2 Glacial deposition

Look at the mark allocation. You don't need to write too much for two marks.

Describe the processes of glacial deposition by completing **boxes 1 and 2** on **Figure 2**.   **(2 marks)**

4   Explain how past UK climate affected the process of creating glaciated upland landscapes.

....

....

....

.... **(4 marks)**

Erosion landforms 1

Guided   1   Study **Figure 1**.

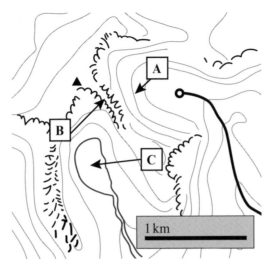

Figure 1 The contour patterns for some glacial landforms

Identify the landforms **A** and **C** on **Figure 1**. Landform B has been named for you.

A ....   B Corrie   C ....   **(2 marks)**

2   Explain the formation of a corrie.

> The formation of a landform is easier to remember if you think about it as a series of stages.

....

....

....

....

....   **(4 marks)**

3   Study **Figure 2**.

Figure 2 An upland glacial landform

(a)   Identify the landform shown in **Figure 2**.

....   **(1 mark)**

(b)   Identify the **two** processes involved in the formation of this landform.

(i) ....   (ii) ....   **(2 marks)**

Erosion landforms 2

1   Study **Figure 1**.

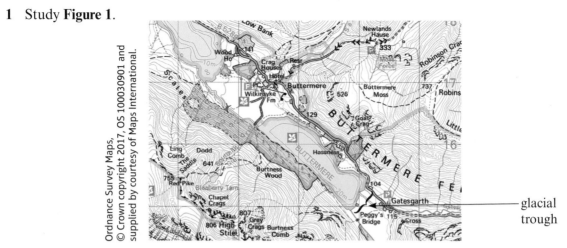

Ordnance Survey Maps,
© Crown copyright 2017, OS 100030901 and
supplied by courtesy of Maps International.

glacial
trough

Figure 1 1:50 000 Ordnance Survey map of Buttermere, the Lake District National Park

Describe how a glacial trough can be recognised on an Ordnance Survey map.

> Landforms are usually recognised by contour patterns.

....

.... **(2 marks)**

2   Study **Figure 2**.

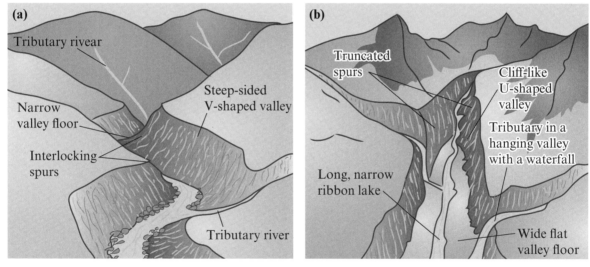

(a)
Tributary rivear

Narrow
valley floor

Interlocking
spurs

Steep-sided
V-shaped valley

Tributary river

(b)
Truncated
spurs

Cliff-like
U-shaped
valley

Tributary in a
hanging valley
with a waterfall

Long, narrow
ribbon lake

Wide flat
valley floor

Figure 2 A valley (a) before, and (b) after, glaciation

(a)   Describe the formation of a hanging valley.

....

.... **(2 marks)**

(b)   Explain how erosional processes form truncated spurs.

Glaciers usually move down a pre-existing river valley. The material transported by

the glacier erodes interlocking spurs ....

....

.... **(4 marks)**

Transport and deposition landforms

1　Study **Figure 1**.

Figure 1 A landform resulting from glacial deposition

(a)　Identify the type of moraine shown at **A** in **Figure 1**.

☐ **A**　Ground　　☐ **B**　Lateral　　☐ **C**　Medial　　☐ **D**　Terminal　　**(1 mark)**

〉**Guided**〉　(b)　Explain how ground moraine is formed.

A glacier transports weathered and eroded material called till. When the glacier melts

.... **(2 marks)**

2　Study **Figure 2**.

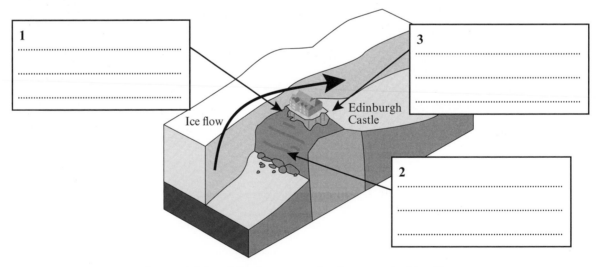

Figure 2 Edinburgh Castle, an example of a crag and tail landform

Explain the formation of a crag and tail landform by adding labels to **boxes 1–3** on **Figure 2**.　　**(4 marks)**

3　Suggest how drumlins would be represented on Ordnance Survey maps.

> Think about the shape of drumlins from above.

....

.... **(2 marks)**

Human activity

1 Explain how water storage and supply developments can change glaciated upland landscapes.

> You do **not** have to write about a specific upland area.

....

....

....

.... **(4 marks)**

Guided 2 Study **Figure 1**.

Figure 1 The Watkin Path, Snowdonia

Suggest **one** advantage and **one** disadvantage of tourism in glaciated uplands.

An advantage of tourism is ....

However, a disadvantage is that footpath erosion can take place ....

.... **(2 marks)**

3 Explain how forestry and farming can change glaciated upland landscapes.

....

....

....

.... **(4 marks)**

⊕ Located example **Glacial development**

1   Identify the rock types found in the Snowdonia
    National Park.

> Rocks in Snowdonia include
> limestone, slate and granite.

☐ **A** Igneous and metamorphic

☐ **B** Only igneous

☐ **C** Metamorphic and sedimentary

☐ **D** Igneous, metamorphic and sedimentary   **(1 mark)**

2   Where do settlements tend to be located in the Snowdonia National Park?

☐ **A** In the valleys

☐ **B** On mountain peaks

☐ **C** On hillsides

☐ **D** There are no settlements in the area   **(1 mark)**

> **Guided**

3   Study **Figure 1**.

Figure 1 Glacial landscape, Snowdonia National Park

Explain how glacial events helped to form Snowdonia's distinctive landscape.

During the last main ice advance, 18 000 years ago, there was an ice cap in the

Snowdonia region. ....

.... **(2 marks)**

4   Examine the most significant factors that have caused change in a named glaciated
    upland landscape.

Named glaciated upland landscape ....

....

.... **(8 marks)**

> Continue your answer on your own paper. You should aim to write approximately one side of A4.

Global atmospheric circulation

1 Study **Figure 1**.

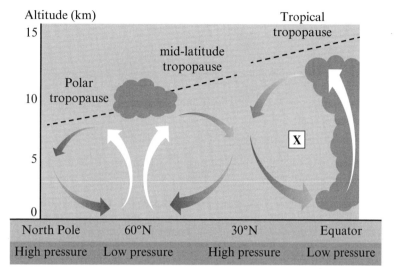

Figure 1 Atmospheric circulation in the northern hemisphere

(a) Name the circulation cell indicated by **X** on **Figure 1**.

     ☐ **A** Equator cell   ☐ **B** Ferrel cell   ☐ **C** Hadley cell   ☐ **D** Polar cell   **(1 mark)**

> **Guided**

(b) Explain how the circulation cell labelled **X** forms.

The highest amount of solar radiation is received at the Equator. This causes warm air

....

....

.... **(4 marks)**

2 Explain how the global atmospheric circulation system transfers heat energy from the Equator to the Poles.

....

....

....

.... **(4 marks)**

3 (a) Identify which **one** of the following is a warm Atlantic current.

     ☐ **A** Canary   ☐ **C** Labrador

     ☐ **B** North Atlantic Drift   ☐ **D** East Greenland   **(1 mark)**

(b) Explain how ocean currents redistribute heat energy across the Earth.

> Remember that you don't have to name ocean currents, but this might help your explanation.

....

....

....

.... **(4 marks)**

31

Had a go ☐ Nearly there ☐ Nailed it! ☐

Natural climate change

1 State the term for the period between major ice advances.

.... **(1 mark)**

2 Study **Figure 1**.

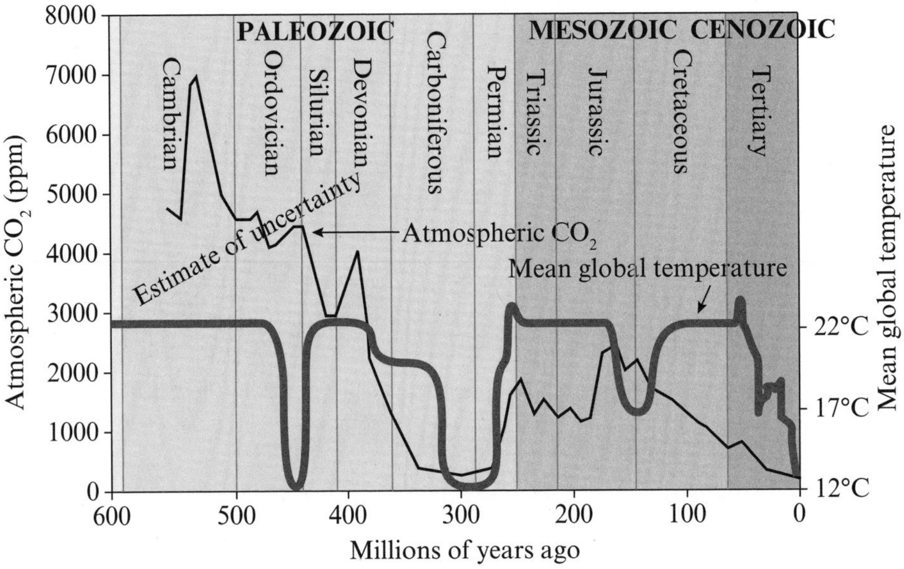

Figure 1 Global temperature variations

From **Figure 1**, calculate how many million years ago the highest global temperatures occurred.

Read graphs carefully. Don't just guess.

.... **(1 mark)**

3 Milankovitch cycles are long-term changes to the Earth's orbit and position that result in changes in climate.

(a) Explain why the eccentricity cycle affects global climate.

....

....

.... **(3 marks)**

(b) State the term for the variation in the tilt of Earth's axis.

.... **(1 mark)**

Guided

4 Explain **two** different sources of evidence which can be used to investigate past natural climate change.

(i) Historical sources, such as diaries. For example, accounts of the 1608 Frost Fair record the Thames freezing over and that ....

....

(ii) ....

.... **(4 marks)**

Human activity

1 Study **Figure 1**.

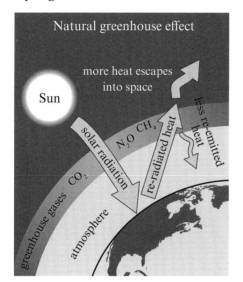 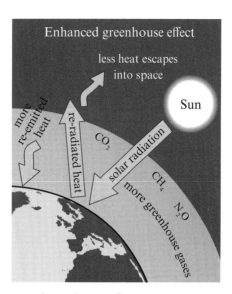

Figure 1 Global climate and greenhouse effects

(a) Define the term **enhanced greenhouse effect**.

.... **(1 mark)**

(b) Compare the natural greenhouse and the enhanced greenhouse effects shown in **Figure 1**.

> **Compare** means that you need to find both similarities and differences.

....

....

....

.... **(4 marks)**

2 (a) Which **one** of the following is **not** a cause of the enhanced greenhouse effect?

     ☐ **A** Industry   ☐ **C** Farming

     ☐ **B** Transport   ☐ **D** Photosynthesis   **(1 mark)**

(b) Explain why transport contributes to the enhanced greenhouse effect.

....

.... **(2 marks)**

Guided 3 Examine the negative impacts of climate change on people and the environment.

Climate change in areas near the Equator, such as Africa's Sahel, is causing less

predictable rainfall and longer periods of drought. This is lowering crop yields

and leading to food shortages and malnutrition. ....

....

.... **(8 marks)**

Continue your answer on your own paper. You should aim to write approximately one side of A4.

Had a go ☐ Nearly there ☐ Nailed it! ☐

The UK's climate

1 (a) State **two** past periods in which the UK had different climates from the UK's climate today.

(i) .... (ii) .... **(2 marks)**

(b) Choose **one** of the periods you named in (a). Explain why the climate was different from the UK today.

....

.... **(2 marks)**

2 Study **Figure 1**.

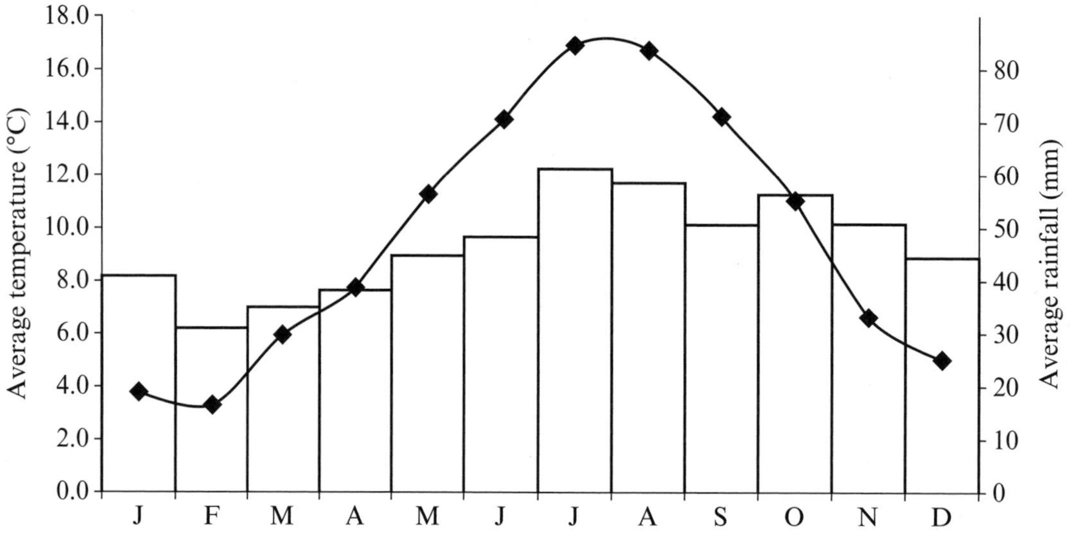

Figure 1 Climate graph for the UK

(a) Calculate the annual temperature range shown in **Figure 1**.

This is the difference in °C between the warmest and coldest months.

.... **(2 marks)**

Guided

(b) Compare the annual precipitation (rainfall) and temperature graphs shown in **Figure 1**.

Both the temperature and the precipitation graphs show a summer maximum.

The highest temperature (17 °C) is in July/August and coincides with the highest precipitation (60 mm). The lowest ....

....

.... **(4 marks)**

3 Examine the significance of the UK's geographical position in relation to its climate.

....

.... **(8 marks)**

Continue your answer on your own paper. You should aim to write approximately one side of A4.

Tropical storms

> **Guided**

1   Study **Figure 1**.

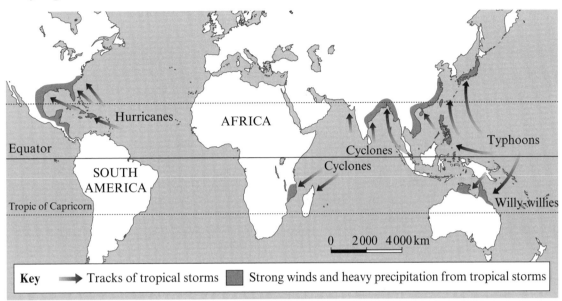

Figure 1 The geographical distribution of tropical cyclones

Explain the distribution of tropical revolving cyclones shown in **Figure 1**.

Tropical cyclones are mostly located in the tropics between 23° north and south

of the Equator. This is because these storms are powered by warm ocean

temperatures – the water needs to be above 26.5 °C and this only occurs in late

summer and autumn in the Tropics. Secondly, ...

....

.... **(4 marks)**

2   State the name for the system that analyses, stores, manipulates and visualises
    geographic information on a map.

    .... **(1 mark)**

3   State the name given to the route taken by a cyclone.

    .... **(1 mark)**

4   Which one of the following is the name given to the centre of a cyclone?

    ☐ **A** Eye wall   ☐ **B** Cloud bank   ☐ **C** Cylinder   ☐ **D** Eye   **(1 mark)**

5   Explain the sequence of the formation
    of a tropical cyclone.

    | A question like this is easier to answer if you write about a series of stages. |

    ....

    ....

    ....

    .... **(4 marks)**

Tropical cyclone hazards

1 (a) Identify the name of the scale used to categorise tropical cyclones.

 ☐ **A** Saffir-Simpson

 ☐ **B** Safir-Simpson

 ☐ **C** Saffir-Simson

 ☐ **D** Saffer-Simpson **(1 mark)**

(b) State the **two** types of information used to categorise tropical cyclones.

 (i) .... (ii) .... **(2 marks)**

(c) Identify the category of tropical cyclone that causes structural damage to buildings.

 ☐ **A** 1

 ☐ **B** 2

 ☐ **C** 3

 ☐ **D** 4 **(1 mark)**

2 Identify **one** natural weather hazard associated with tropical cyclones.

 ☐ **A** River flooding

 ☐ **B** Snowstorms

 ☐ **C** Landslides

 ☐ **D** Drought **(1 mark)**

3 Explain why tropical cyclones cause the following weather hazards.

(a) High winds

....

....

.... **(3 marks)**

(b) Intense rainfall

....

....

.... **(3 marks)**

> Guided

(c) Coastal flooding

Tropical cyclones can cause a large mass of water to reach coastal areas.

Reduced pressure at the centre of the cyclone ....

.... **(3 marks)**

⊕ Located example **Hurricane Sandy**

1   Study **Figure 1**.

Rank	Hurricane	Damage cost to nearest $ billion
1	Katrina	108.0
2	Sandy	71.0
3	Ike	30.0
4	Andrew	27.0
5	Wilma	21.0
6	Ivan	19.0
7	Irene	16.0
8	Charley	15.0
9	Rita	12.0
10	Frances	10.0

Figure 1 The cost of ten USA Atlantic coast tropical storms

Calculate the median cost of the USA Atlantic tropical storms shown in **Figure 1**. Show your calculations.

....
....
....
....
....
.... **(2 marks)**

> If there is an even number of ranked values, the median is the mean of the two central numbers.

2   Suggest why Hurricane Sandy was one of the most expensive Atlantic tropical storms to have affected the USA.

....
....
....
.... **(4 marks)**

3   Name **one** environmental impact of Hurricane Sandy?

.... **(1 mark)**

4   Name **one** social media method used to assess the impacts of Hurricane Sandy.

.... **(1 mark)**

> **Guided**

5   Explain the social impacts of a tropical cyclone on a named developed country.

Named developed country   USA

The worst social impact of Hurricane Sandy was the number of deaths. 150 people were killed, which caused immense trauma to families in the affected regions.

....
.... **(4 marks)**

37

🌐 Located example **Typhoon Haiyan**

1   Study **Figure 1**.

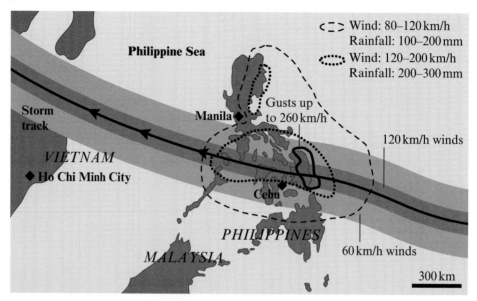

Figure 1 Typhoon Haiyan, November 2013

(a)   State the category of Typhoon Haiyan on the Saffir-Simpson scale.

.... **(1 mark)**

(b)   State the compass direction of the track of Typhoon Haiyan.

.... **(1 mark)**

(c)   Identify the country most affected by Typhoon Haiyan.

    ☐ **A** The Philippines   ☐ **B** Malaysia   ☐ **C** Vietnam   ☐ **D** Laos   **(1 mark)**

Guided   2   Explain **one** environmental impact of Typhoon Haiyan.

Mangrove trees were damaged and uprooted across the islands. This disrupted

the fragile ecosystems such as ....

By trapping nutrients and sediments from drainage, mangroves protect

....

.... **(2 marks)**

3   Evaluate the following statement. 'Responses to tropical cyclones are more effective in developed countries than in developing countries.'

> **Evaluate** means that you need to judge the importance of a factor. Here you need to think about the importance of the level of development in dealing with the effects of cyclones.

....

....

....

.... **(8 marks)**

> Continue your answer on your own paper. You should aim to write approximately one side of A4.

Drought causes and locations

> **Guided**   **1** Compare **two** characteristics of arid and drought environments.

Arid environments have a permanent low precipitation – usually 10–250 mm a year – whereas drought conditions are caused by a temporary period of low precipitation. Both ....

.... **(4 marks)**

2 (a) Identify **one** natural cause of drought.

.... **(1 mark)**

(b) Explain **one** reason for the natural cause you identified in (a).

....

.... **(2 marks)**

3 Explain why global circulation makes some locations more likely to be affected by drought than others.

....

....

....

.... **(4 marks)**

4 Study **Figure 1**.

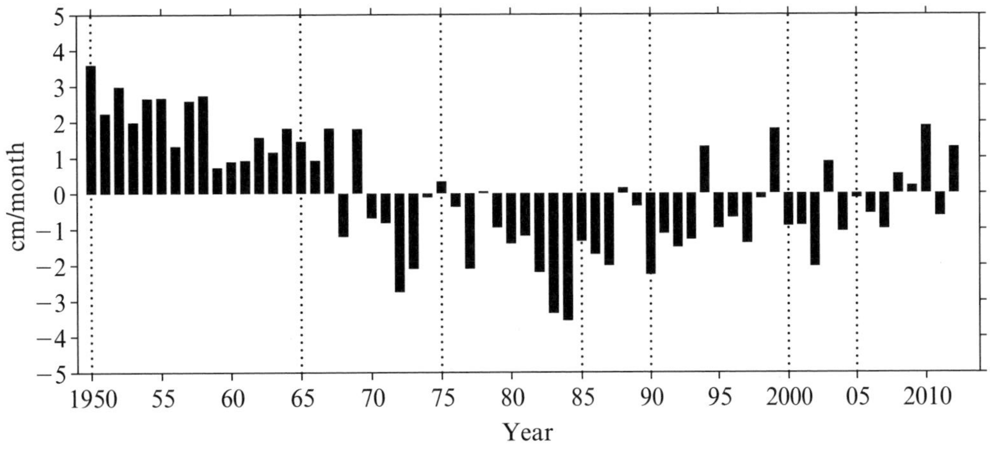

Figure 1 Medium term rainfall trends for the Sahel region of Africa

(a) Calculate the rainfall range of the data shown in **Figure 1**.

.... **(1 mark)**

(b) Describe the variation in rainfall trend between 1990 and 2010 shown in **Figure 1**.

> Always include data when you are describing graphs.

....

....

.... **(3 marks)**

🌐 Located example California, USA

1   Explain reasons why droughts in California can be hazardous.

....

....

....   **(4 marks)**

2   Study **Figure 1**.

Explain **two** impacts of drought on ecosystems in California. Use the information in **Figure 1** in your answer.

DROUGHT IN CALIFORNIA

A large blue-green algal bloom has developed in the Sacramento-San Joaquin Delta. This type of algae produces toxins that, in high concentrations, are lethal to fish and people.

Figure 1 An extract from a Californian newspaper article

....

....

....   **(4 marks)**

3   Explain **two** impacts of drought on people in California.

....

....

....   **(4 marks)**

4   Study **Figure 2**.

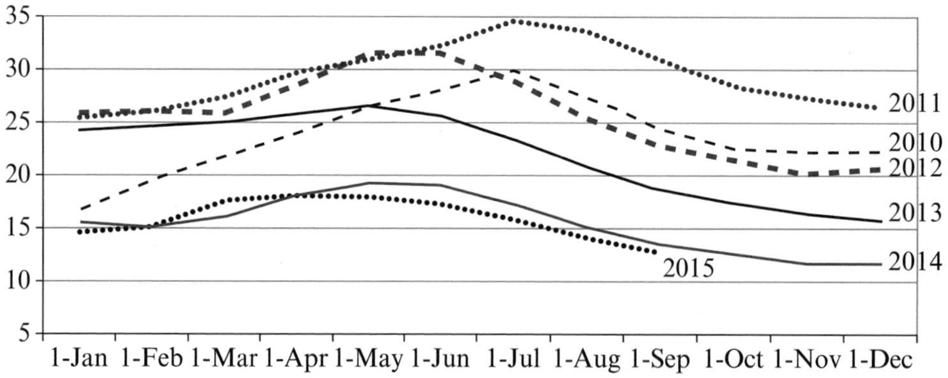

Figure 2 California reservoir storage, million acre feet (one acre foot equals 1 481 349 litres)

Describe the variations in reservoir storage between 2010 and 2015.

> Describe general trends first.

....

....

....   **(3 marks)**

40

⊕ Located example **Ethiopia**

1   Study **Figure 1**.

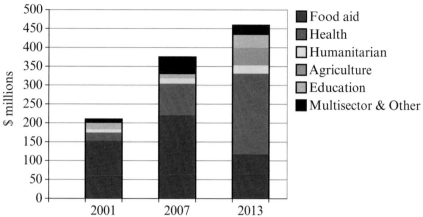

Figure 1 US aid to Ethopia

Guided   Describe the variations in food aid between 2001 and 2013 shown in **Figure 1**.

> Include data in your answer.

The overall pattern for food aid is an increase on the 2001 figures in 2007 but

....

.... **(3 marks)**

2   Suggest possible impacts of severe droughts, such as the drought of 2015, on ecosystems in Ethiopia.

> This focuses on Ethiopia, but you can apply your wider knowledge to complete the answer.

....

....

....

.... **(4 marks)**

3   Compare the different responses to drought in a named developed and in a named emerging or developing country.

> Remember to use the located examples that you have done in class.

Named developed country ....

Named emerging or developing country ....

....

....

....

.... **(8 marks)**

> Continue your answer on your own paper. You should aim to write approximately one side of A4.

The world's ecosystems

1 (a) State **two** characteristics of tundra ecosystems.

(i) ....   (ii) .... **(2 marks)**

(b) Which **one** of the following is an area where large-scale desert ecosystems are **not** located?

> Read questions carefully. This one is asking where desert ecosystems are **not** located.

☐ **A** North Africa   ☐ **C** South-west North America

☐ **B** Central Australia   ☐ **D** Central South America   **(1 mark)**

Guided

2 Study **Figure 1** and **Figure 2**.

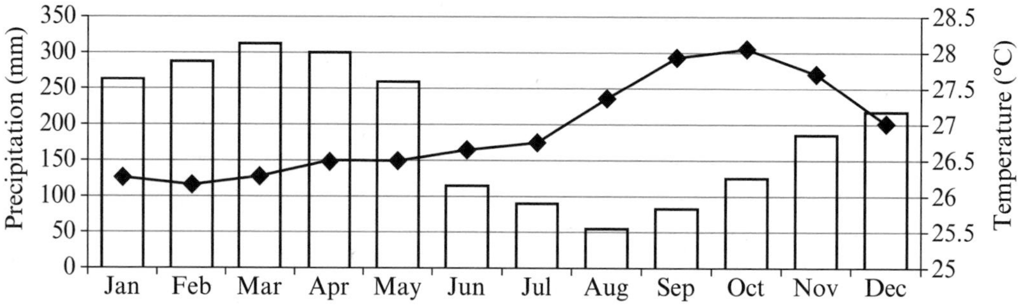

Figure 1 Tropical forest climate graph

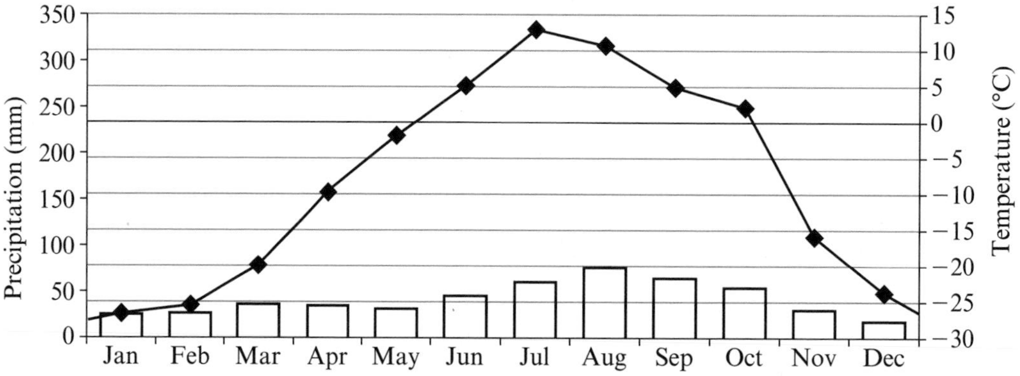

Figure 2 Boreal forest climate graph

Compare the climate graphs for the tropical and boreal forest ecosystems.

Temperatures for the tropical forest graph are above 26 °C all year and rise to a

maximum of 27.5 °C in October. However, temperatures in the boreal forest only

reach above freezing between May and ....

....

.... **(3 marks)**

3 Describe **one** way that climate influences the distribution of large-scale ecosystems.

....

.... **(2 marks)**

4 Name **one** local factor which affects ecosystems.

.... **(1 mark)**

Importance of the biosphere

1   State the correct term for the layer of the Earth where life exists.

....   **(1 mark)**

2   Identify **two** building materials provided by the biosphere.

(i) ....   (ii) ....   **(2 marks)**

3   Explain how the biosphere provides **one** fuel resource.

....

....   **(2 marks)**

4   Study **Figure 1**.

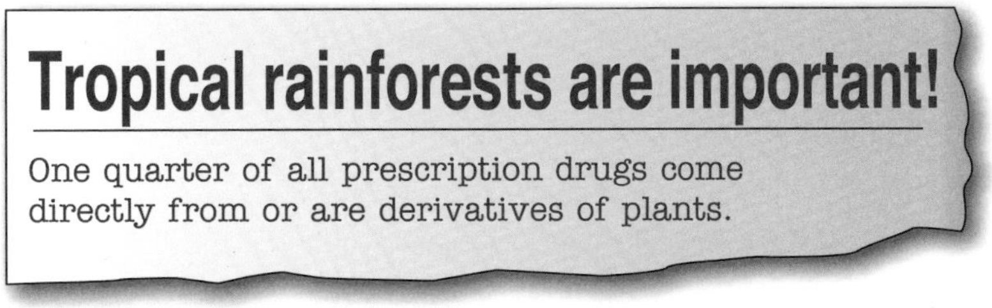

Tropical rainforests are important!

One quarter of all prescription drugs come
directly from or are derivatives of plants.

Figure 1 A newspaper extract about tropical forests

Explain how the biosphere provides medicines for people. Use the information in
Figure 1 as part of your answer.

....

....

....   **(4 marks)**

5   Study **Figure 2**.

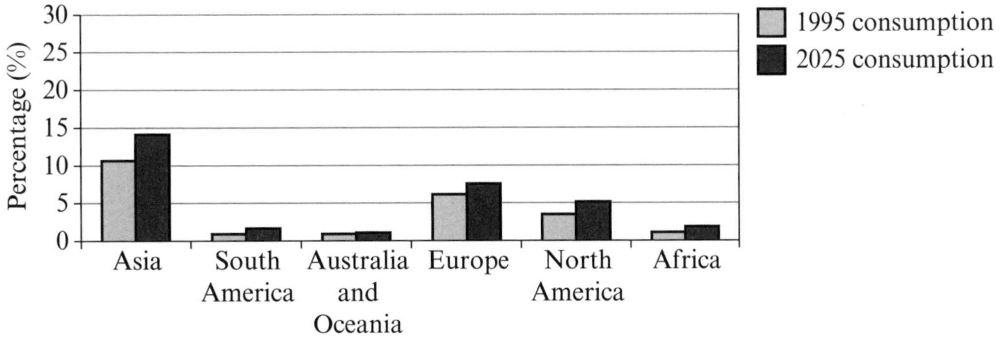

Figure 2 Actual and projected global water consumption

(a)   From **Figure 2**, calculate the increase in water
      consumption between 1995 and 2025 for Asia.   Give your answer as a percentage.

....   **(1 mark)**

(b)   Explain **one** problem caused by the increased exploitation of water resources.

....

....   **(2 marks)**

The UK's main ecosystems

> **Guided**

1 State **two** characteristics of moorlands.

(i) ....   (ii) Found in upland areas   **(2 marks)**

2 Which **two** of the following are characteristics of wetlands?

☐ **A** Dry soils

☐ **B** Waterlogged soils

☐ **C** Coniferous trees

☐ **D** Rich in nutrients

☐ **E** Sandy soils

☐ **F** Low in nutrients   **(2 marks)**

3 Study **Figure 1**.

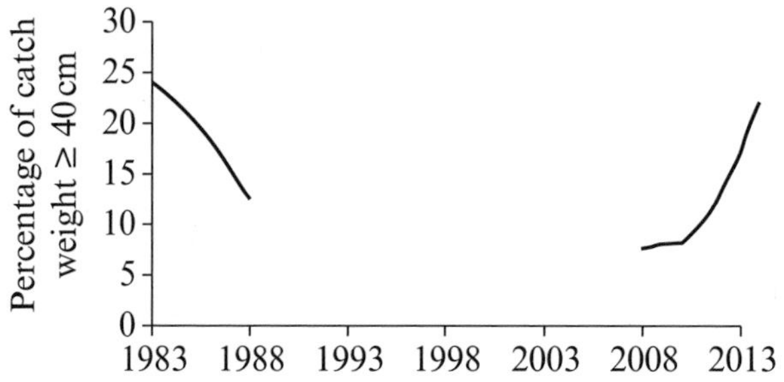

Figure 1 Large fish caught (40 cm long plus), by weight, in the north-western North Sea, 1983 to 2014

(a) Plot the data in the table to complete the line graph on **Figure 1**.

Year	Percentage of catch weight ≥ 40 cm
1993	7
1998	6
2003	7

(3 marks)

(b) Suggest **one** reason for the low percentage of fish caught between 1988 and 2008.

....   **(1 mark)**

(c) Explain how human activities are degrading marine ecosystems.

> Make sure that you understand specification terms. **Degrading** means making worse.

....

....

....

....   **(4 marks)**

Tropical rainforest features

1 Study **Figure 1**.

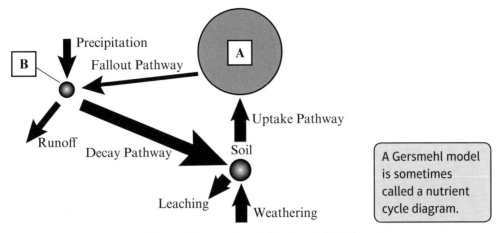

Figure 1 Gersmehl model for tropical rainforests

> A Gersmehl model is sometimes called a nutrient cycle diagram.

(a) Name the nutrient stores labelled **A** and **B** on **Figure 1**.

A: ....   B: .... **(2 marks)**

(b) Explain why nutrient store **A** is the largest store.

....

.... **(2 marks)**

> **Guided**

(c) Explain **two** reasons why tropical rainforest soils are nutrient-poor.

(i) The nutrients in the soil are rapidly taken up by plants as nutrients are needed

for the high rates of growth and photosynthesis.

(ii) ....

.... **(4 marks)**

2 Which **one** of the following is a biotic characteristic of rainforests?

☐ **A** Soil   ☐ **C** Water

☐ **B** Plants   ☐ **D** Climate   **(1 mark)**

3 (a) State the type of weathering that occurs in tropical rainforests.

.... **(1 mark)**

(b) Explain why this type of weathering occurs.

....

.... **(2 marks)**

4 Explain the contribution made by humans to the functioning of the tropical rainforests.

....

.... **(2 marks)**

TRF biodiversity and adaptations

1   Identify which **one** of the following is **not** an animal adaptation for living in tropical rainforests.

☐ **A** Strong limbs

☐ **B** Large percentage of body fat

☐ **C** Camouflage

☐ **D** Adapted beaks   **(1 mark)**

2   Identify the type of trees found in tropical rainforests.

☐ **A** Evergreens

☐ **B** Conifers

☐ **C** Boreal

☐ **D** Deciduous   **(1 mark)**

> Take time to think about each option.

Guided   3   Study **Figure 1**.

1 Emergent trees ....
....
....
....
....
....

2 Lianas are woody vines that climb up tree trunks so that ....
....
....
....

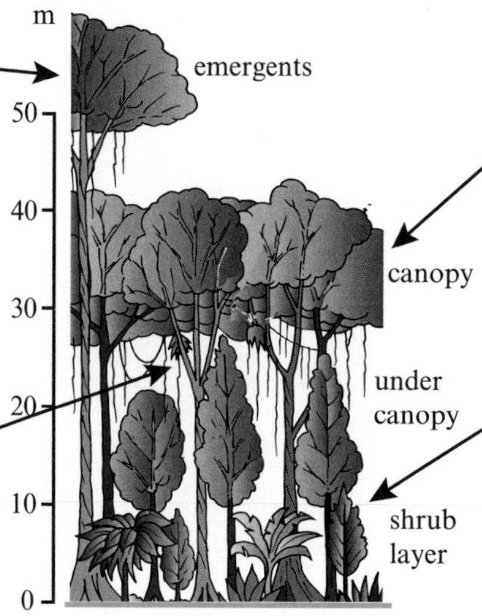

3 Many plants have modified leaves with drip-tips, which
....
....
....
....

4 The tallest trees often have buttress roots, which
....
....
....
....

Figure 1 Vegetation layers in a tropical rainforest

Explain how plants are adapted to the tropical rainforest environment by completing **boxes 1–4** on **Figure 1**.   **(4 marks)**

4   Define the term **biodiversity.**

....   **(1 mark)**

5   Explain why tropical rainforests have very high biodiversity.

....

....

....

....   **(4 marks)**

TRF goods and services

1 Which **one** of the following is a service provided by tropical rainforests?

☐ **A** Producing CO_2 ☐ **C** An O_2 store

☐ **B** Home to indigenous tribes   ☐ **D** Low biodiversity   **(1 mark)**

2 State **two** goods provided by tropical rainforests.

(i) .... (ii) .... **(2 marks)**

3 Study **Figure 1**.

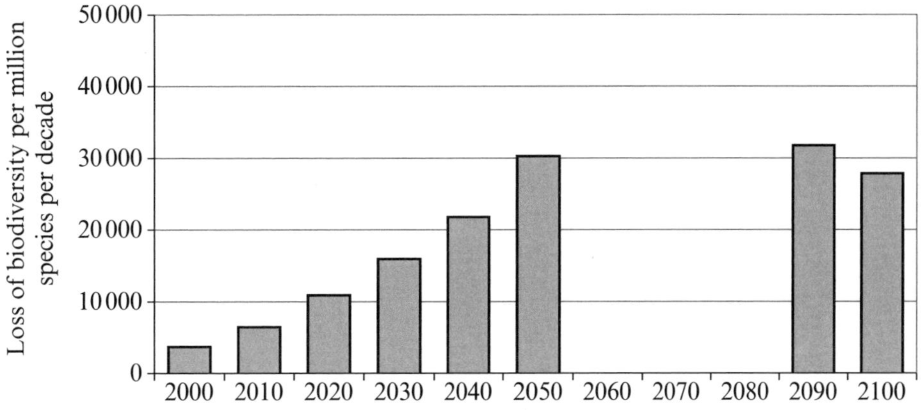

Figure 1 Actual and projected changes of biodiversity in rainforest forests, 2000–2100

Decade	Projected extinctions per million species per decade
2060	48 000
2070	42 000
2080	38 000

Figure 2 Table of data on projected extinctions

(a) Plot the data in the table on to **Figure 2** to complete **Figure 1**.

> Use a ruler so that the tops of the bars are straight.

(3 marks)

(b) Suggest reasons for this loss of biodiversity.

....

....

.... **(3 marks)**

Guided

4 Suggest why the structure of tropical rainforests might change.

Climate change may result in reduced precipitation and long periods of

drier conditions. This will .... the biomass store.

Therefore the layered structure of the rainforest will be reduced because

....

.... **(4 marks)**

Deforestation in tropical rainforests

1 Explain why resource extraction is causing the deforestation of tropical rainforests.

....

.... **(2 marks)**

2 Study **Figure 1**.

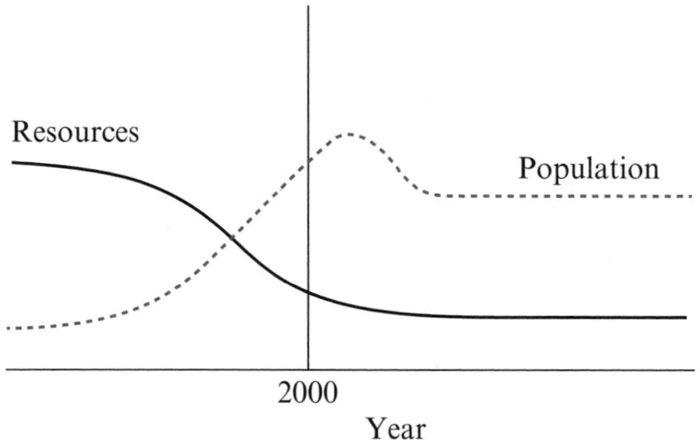

Figure 1 Actual and projected global population and available resources

Compare the line graphs for global population and available resources.

....

....

.... **(3 marks)**

3 Study **Figure 2**.

It is estimated that 35–45% of total rainforest deforestation is for small-scale agriculture, 20–25% for cattle ranching and 15–20% for large-scale agriculture, such as plantations.

Figure 2 An extract from a newspaper article about rainforest deforestation

(a) According to **Figure 2**, which **one** of the following is the **maximum** percentage of total rainforest deforestation for agriculture?

☐ **A** 75   ☐ **B** 65   ☐ **C** 90   ☐ **D** 70   **(1 mark)**

⟩ **Guided** ⟩ (b) Explain **one** reason why tropical rainforests are being deforested for agriculture.

Deforestation is taking place for large-scale agriculture such as oil palm plantations.

There is a high global demand for palm oil and therefore ....

.... **(2 marks)**

4 State **one** social cause of rainforest deforestation.

Make sure you know the differences between economic and social causes of deforestation.

.... **(1 mark)**

🌐 Located example Tropical rainforest management

1 Define the term **sustainable management**.

.... **(1 mark)**

2 Study **Figure 1**.

> **The value of rainforests?**
> * This is controversial
> * Researchers not able to give figures
> * Value of future cures for diseases, ecotourism and hydropower is not known

Figure 1 A slide from a presentation about tropical rainforest management

Guided

(a) Define the term **ecotourism**.

This is responsible tourism to unspoilt areas that preserves ....

.... **(1 mark)**

(b) Explain **one** way that ecotourism contributes to the sustainable management of the Amazon rainforest.

....

.... **(2 marks)**

3 Explain **one** way that commodity value is contributing to the sustainable management of the Amazon rainforest.

> Governmental and regional authority policies can be called governance.

....

.... **(2 marks)**

4 Identify **one** policy which is not a government policy for the sustainable management of the Amazon rainforest.

.... **(1 mark)**

5 Explain the advantages of using reduced impact logging in the sustainable management of the Amazon rainforest.

....

....

....

.... **(4 marks)**

Deciduous woodlands features

1 In which **three** of the following countries can deciduous woodlands **mainly** be found?

> Deciduous woodlands are found in more than one country. Look at the marks available.

☐ **A** Brazil   ☐ **C** Japan

☐ **B** The UK   ☐ **D** China

(3 marks)

Guided

2 Study **Figure 1**.

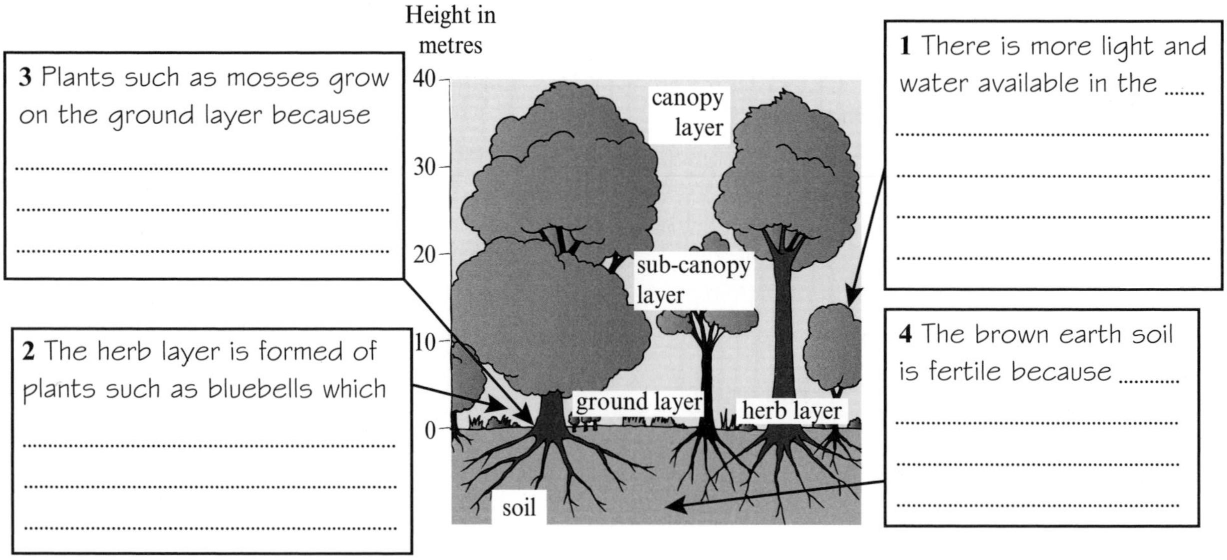

3 Plants such as mosses grow on the ground layer because
....
....
....

1 There is more light and water available in the ....
....
....
....
....

2 The herb layer is formed of plants such as bluebells which
....
....
....

4 The brown earth soil is fertile because ....
....
....
....

Height in metres

canopy layer

sub-canopy layer

ground layer   herb layer

soil

Figure 1 The structure and functioning of deciduous woodland

(a) Identify (i) **one** biotic and (ii) **one** abiotic characteristic of deciduous woodlands shown on **Figure 1**.

(i) ....   (ii) ....   **(2 marks)**

(b) Explain the functioning of deciduous woodlands by completing **boxes 1–4** on **Figure 1**.   **(4 marks)**

3 Identify **two** characteristics of a typical deciduous woodland climate.

> Remember that the annual temperature range is the difference between the highest and lowest monthly temperatures.

☐ **A** Precipitation all year with a summer maximum

☐ **B** Precipitation all year with a winter maximum

☐ **C** An annual temperature range of approximately 12 °C

☐ **D** An annual temperature range of approximately 20 °C   **(2 marks)**

4 Explain why the Gersmehl model (nutrient cycle model) for deciduous woodlands has biomass and soil nutrient stores that are similar in size.

....
....
....
....   **(4 marks)**

Deciduous woodlands adaptations

1   Study **Figure 1**.

Ecosystem type	Net primary productivity (kilocalories/square metre/year)	Approximate kilocalories per square metre per day
Tropical forest	9000	25
Tropical grassland	3000	8
Temperate grassland	2000	6
Deciduous woodlands	6000	16
Boreal forest	3500	10

Figure 1 Net primary production for selected large-scale ecosystems

(a) Calculate the mean net primary production for the five ecosystems shown in **Figure 1**.

.... **(1 mark)**

(b) Explain **one** reason why deciduous woodlands have above the mean net primary production.

....

.... **(2 marks)**

2   Explain **two** ways that animals have adapted to living in deciduous woodland environments.

....

....

....

.... **(4 marks)**

> **Guided**

3   Study **Figure 2**.

1 Deciduous leaves are

....

This enables them to carry out maximum photosynthesis

....

2 Deciduous trees drop their leaves in autumn because

....

....

Figure 2 A leaf from a deciduous oak tree

Explain how leaves are adapted to the deciduous woodland environment by completing **boxes 1** and **2** on **Figure 2**.   **(2 marks)**

4   State **one** reason why deciduous trees have extensive, deep root systems.

> As the command word is **state**, you do not have to add any explanation.

.... **(1 mark)**

Deciduous woodlands goods and services

1 Explain the provision of **one** service by deciduous woodlands.

....

.... **(2 marks)**

2 Identify **two** goods provided by deciduous woodlands.

☐ **A** Arable crops   ☐ **C** Oxygen

☐ **B** Fuel   ☐ **D** Timber

> Make sure you know the difference between goods and services.

(2 marks)

> **Guided**

3 Suggest how climate change might affect the biodiversity of deciduous woodlands.

Climate change may result in drier summers. Trees affected by drought may die,

which would lower biodiversity as insects and ....

Biodiversity could also be reduced because ....

.... **(4 marks)**

4 Study **Figure 1**.

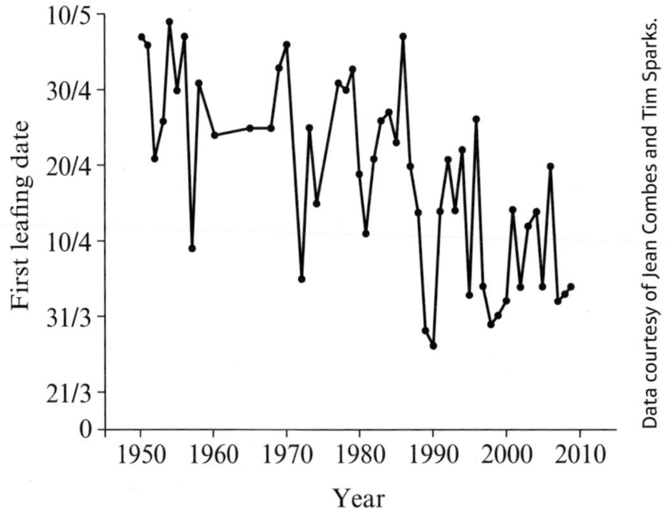

Figure 1 Date of the appearance of the first leaves on oak trees between 1950 and 2008

(a) State the general trend for the first appearance of leaves shown in **Figure 1**.

.... **(1 mark)**

(b) Suggest reasons for this trend. Use evidence from **Figure 1** in your answer.

....

....

....

.... **(4 marks)**

Deforestation in deciduous woodlands

1 Study **Figure 1**.

> The Domesday Book records that approximately 15 per cent of England was once covered by woodland. By the end of the 19th century, woodland cover had dropped to below 5 per cent. Since then England's forest and woodland areas have been expanding, and by the beginning of the 21st century there were over 1.1 million hectares or 8.4 per cent woodland cover. However, this is relatively low compared to Europe, which has 44 per cent. In 2016, the UK government supported plans to create a new national forest in England. (*Note: not all woodlands referred to are deciduous*)

Figure 1 An extract from a report on the decline of woodland areas in England

> **Guided**

(a) Use the information from **Figure 1** to construct the bars for the 19th and 21st centuries in the graph below.

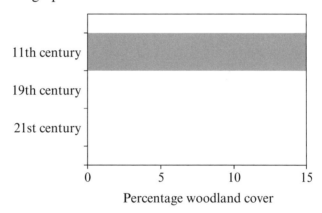

Percentage woodland cover

(2 marks)

(b) State **one** social cause for the reduction in woodlands between the 11th and 19th centuries.

> It's important to know the differences between social and economic causes of deforestation.

.... **(1 mark)**

(c) Explain why this led to the deforestation of woodlands.

....

.... **(2 marks)**

2 Which **two** of the following are reasons why agricultural change has caused the deforestation of woodland areas?

☐ **A** More farmland needed   ☐ **C** Increased use of pesticides and herbicides

☐ **B** Destruction by cattle   ☐ **D** Present government policy   **(2 marks)**

3 Explain **one** disadvantage of replacing deciduous woodlands with coniferous trees.

....

.... **(2 marks)**

🌐 **Located example** Deciduous woodlands management

1   Explain why deciduous woodland areas require sustainable management.

   ....

   ....

   ....

   ....   **(4 marks)**

2   Identify **one** use of deciduous woodlands.

   ☐ **A** Plantation agriculture

   ☐ **B** Tourism

   ☐ **C** Urbanisation

   ☐ **D** Manufacturing industry   **(1 mark)**

▷ **Guided** ▷ 3   Suggest how **two** different policies make the management of deciduous woodlands more sustainable.

   Careful management, with dedicated walking and cycling routes in more fragile areas,

   means that ....

   Landowners are given grants to plant native tree species, which means ....

   ....

   ....   **(4 marks)**

4   'Management of tropical rainforests is more complex than the management of deciduous woodlands.'
   Assess this statement.

   > Although the question doesn't ask for named regions, you should use information about the specific regions you have studied.

   ....

   ....

   ....

   ....

   ....

   ....   **(8 marks)**

   > Continue your answer on your own paper. You should aim to write approximately one side of A4.

Paper 1

In this question, four additional marks will be awarded for your spelling, punctuation and grammar and for your use of specialist terminology.

Assess the relative importance of economic and social causes for the deforestation of tropical rainforests.

For this question you need to consider both **economic** causes (conversion to agriculture, resource extraction) and **social** causes (population pressure). Then use the information to decide and explain which of the causes is more important.

The content for Assessment Objective 2, which is worth 4 marks, might include the following explanations for deforestation.

- Tropical rainforests are being cleared to make room for cattle ranching due to increasing demands for beef.
- Extensive areas are being cleared for plantations, particularly palm oil. There is a high global demand for palm oil as it is used in food products and cosmetics, and as biofuel.
- Important mineral resources are often located in rainforest areas. Demand from developed countries make it economically profitable to extract these minerals.
- Logging involves timber companies cutting down mature trees such as mahogany and teak and selling the wood to developed countries.
- Logging, agriculture and mineral extraction require infrastructure such as roads, increasing the amount of deforestation.
- Governments and international aid agencies have encouraged people to move into rainforests to help reduce poverty in urban areas.

For Assessment Objective 3, which is worth 4 marks, you would need to:

- **weigh up** the relative importance of both the social and economic factors causing the deforestation of rainforests
- **make a judgement** based on evidence that includes facts and figures.

For SPGST, which is worth 4 marks, you need to:

- **spell** (especially geographical terms) and punctuate accurately
- make sure you follow the rules of **grammar**
- use a wide range of **geographical terms**.

....

....

....

....

....

....

(8 marks plus 4 marks for SPGST)

Continue your answer on your own paper. You should aim to write about a side of A4.

Had a go ☐ Nearly there ☐ Nailed it! ☐

An urban world

1 Define the term **urbanisation**.

.... **(1 mark)**

2 Study **Figure 1**.

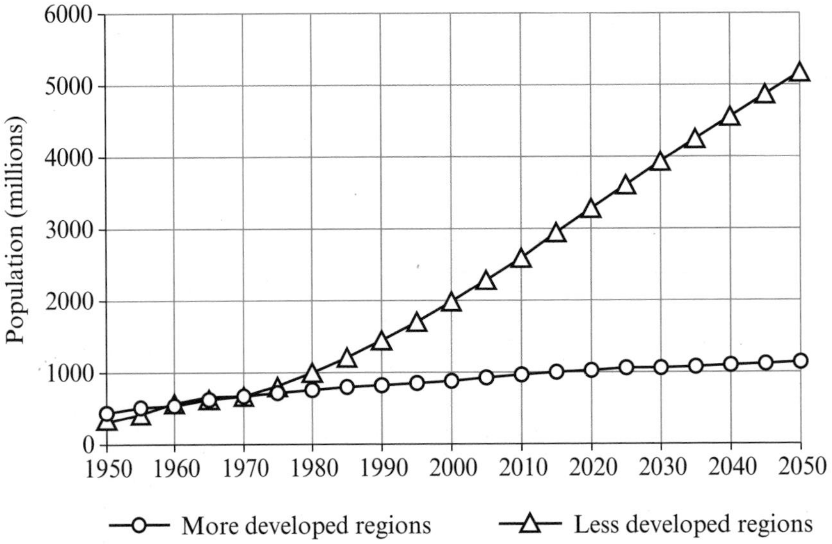

Figure 1 Actual and projected numbers of people living in urban areas in developed and less developed regions

Guided

(a) From **Figure 1**, compare the graphs for the developed and less developed regions.

Both the graphs show an increase in the number of people living in urban areas. However, the graph for developed regions increases much more slowly than the developing graph, rising from 400 million in 1950 to a projected figure of 1100 million in 2050. The developing regions graph ....

....

.... **(4 marks)**

(b) Suggest **one** reason for urbanisation in developed countries.

....

.... **(2 marks)**

3 State the term for unplanned settlements found in urban areas in developing countries.

.... **(1 mark)**

4 Explain **two** problems caused by rapid urbanisation in developing countries.

Make sure you know the different effects of urbanisation in developed, developing and emerging countries.

....

....

....

.... **(4 marks)**

UK urbanisation differences

Guided   1   Study **Figure 1**.

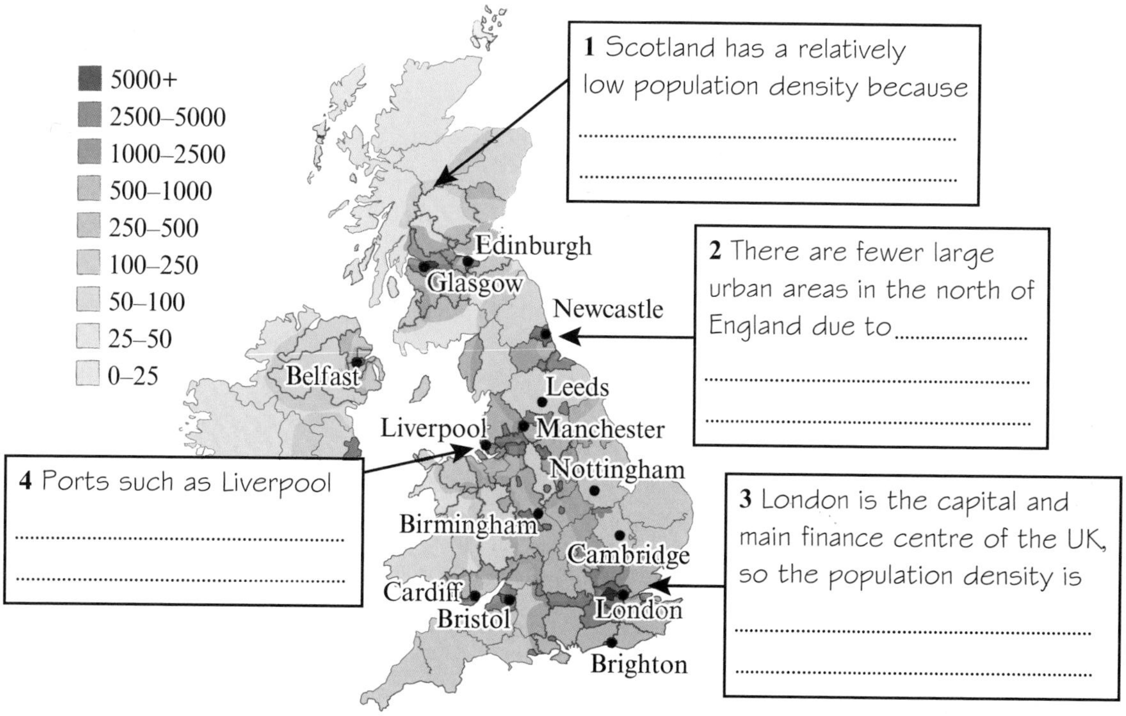

Legend:
- 5000+
- 2500–5000
- 1000–2500
- 500–1000
- 250–500
- 100–250
- 50–100
- 25–50
- 0–25

Map labels: Edinburgh, Glasgow, Newcastle, Belfast, Leeds, Liverpool, Manchester, Nottingham, Birmingham, Cambridge, Cardiff, Bristol, London, Brighton

1 Scotland has a relatively low population density because
....
....

2 There are fewer large urban areas in the north of England due to ....
....
....

4 Ports such as Liverpool
....
....

3 London is the capital and main finance centre of the UK, so the population density is
....
....

Figure 1 The population density of the UK

(a)   Explain the distribution of the UK's population by completing **boxes 1–4** on **Figure 1**.   **(4 marks)**

(b)   Identify the mean population density for the UK.

☐ **A**   166 people per km^2

☐ **B**   366 people per km^2

☐ **C**   206 people per km^2

☐ **D**   266 people per km^2 **(1 mark)**

2   Identify which **two** of the following are major urban centres in the UK.

☐ **A**   Birmingham

☐ **B**   Stroud

☐ **C**   Leeds

☐ **D**   Malham   **(2 marks)**

3   Explain **two** factors that cause the degree of urbanisation to vary between regions of the UK.

> The degree of urbanisation means how many people live in towns and cities as opposed to rural areas.

....
....
....
....   **(4 marks)**

57

🌐 Case study **Context and structure**

> You will have studied a named major city in the UK. As you work through the following pages, answer using your own case study city where possible.

1 Explain the connectivity of a named UK city.

> **Connectivity** means how easily a city can be reached from other places.

Named UK city ....

....

....

....

.... **(4 marks)**

2 Define the term **site** in relation to a settlement.

.... **(1 mark)**

Guided   **3** State the meaning of the term **situation** of a settlement.

A settlement's location in relation to surrounding .... features   **(1 mark)**

4 Study **Figure 1**.

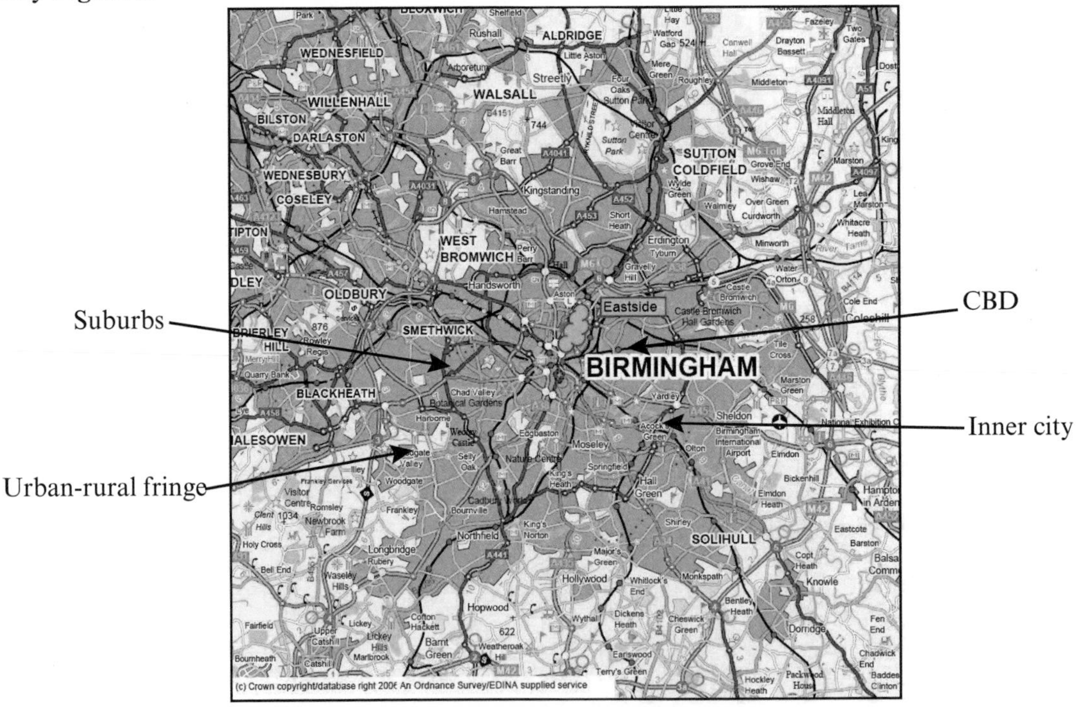

Figure 1 The structure of some areas of Birmingham

Explain the structure of a named UK city. Use the information in **Figure 1** to help you.

Named UK city ....

....

....

....

.... **(4 marks)**

🌐 Case study **A changing UK city**

1   Study **Figure 1**.

| 1 Urbanisation | → 2 .... | → 3 Counter-urbanisation | → 3 Re-urbanisation |

Figure 1 The sequence of process change in an urban centre

(a)   Fill in the correct term for **process 2** shown on **Figure 1**.

.... **(1 mark)**

(b)   Define the term **counter-urbanisation**.

.... **(1 mark)**

Guided (c)   Explain **one** reason why counter-urbanisation takes place.

People move from urban areas because traffic congestion leads to poor air quality,

so .... **(2 marks)**

2   Identify the meaning of the term **re-urbanisation**.

☐ **A** New towns are built   ☐ **C** Extensive suburbs develop

☐ **B** Urban areas expand   ☐ **D** People move back to the centre of urban areas   **(1 mark)**

3   Study **Figure 2**.

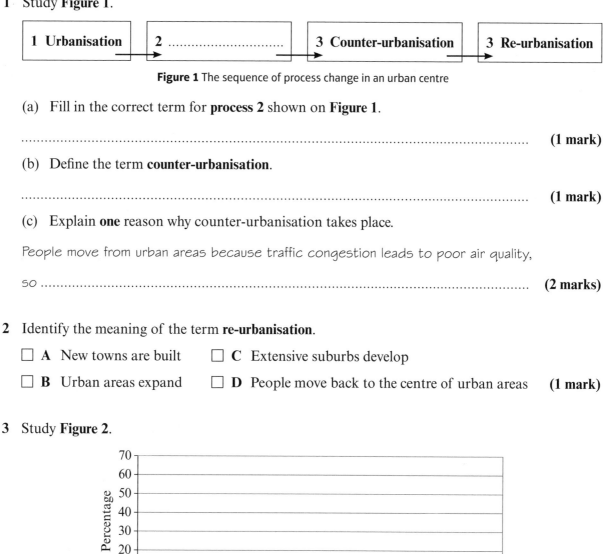

Figure 2 Population of Birmingham by ethnic group, 2011 census

(a)   Plot the data in the table to complete **Figure 2**.

| White British | 53% |
| Indian | 6% |

(2 marks)

(b)   Explain **one** way in which migration affects a named UK city.

> You should use the city you have studied in class.

Named UK city ....

....

.... **(2 marks)**

Globalisation and economic change

1 Study **Figure 1**.

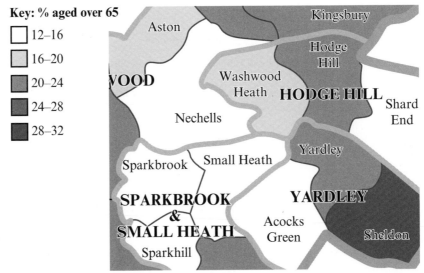

Key: % aged over 65
- ☐ 12–16
- 16–20
- 20–24
- 24–28
- 28–32

Figure 1 Percentage of the population aged over 65 for selected wards in Birmingham

Shade in the Shard End and Acocks Green areas using the information in the table below.

Ward name	Population aged over 65 (%)
Shard End	28–32
Acocks Green	20–24

> Look carefully at the key to find the correct shading.

2 Explain **one** reason for the change in population size of a named city.

Named UK city ....

....

.... **(2 marks)**

> **Guided**

3 Explain **one** cause of decentralisation in a named city.

> This focuses on Birmingham, but you can apply your wider knowledge to complete the answer.

The NEC (National Exhibition Centre) was built near junction 6 of the M42 motorway,

east Birmingham, to encourage ....

.... **(2 marks)**

4 State the definition of the term **deindustrialisation**.

.... **(1 mark)**

5 Assess the impacts of deindustrialisation on a named city.

Named UK city ....

....

.... **(8 marks)**

> Continue your answer on your own paper. You should aim to write approximately one side of A4.

🌐 Case study City inequalities

1   Study **Figure 1**.

☐ in the most deprived 5% in England
▨ in the most deprived 10% but not the most deprived 5%
■ in the most deprived 25% but not the most deprived 10%
☐ not in the most deprived 25% and non-residential areas

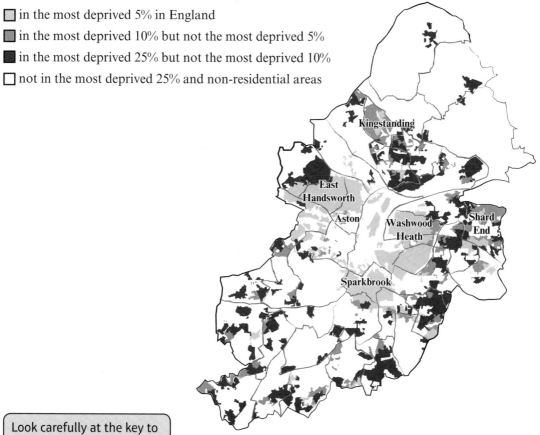

Figure 1 Areas of deprivation in Birmingham

> Look carefully at the key to identify the correct areas.

Using **Figure 1**, describe the distribution in Birmingham of areas that are in the most deprived 5% in England.

....

....

.... **(3 marks)**

2   Suggest **one** reason why economic change is increasing inequality in a named city.

Named UK city ....

....

.... **(2 marks)**

3   Explain why there are differences in the quality of life in a named city.

....

....

....

.... **(4 marks)**

Had a go ☐ Nearly there ☐ Nailed it! ☐

Case study Retailing changes

1 Explain **two** reasons why internet shopping is becoming more popular.

....

....

....

.... **(4 marks)**

2 Which **one** of the following is the **main** function of an urban CBD?

☐ **A** Residential

☐ **B** Manufacturing

☐ **C** Recreation

☐ **D** Retail and offices **(1 mark)**

> Make sure you know abbreviations of key terms, for example, CBD.

3 (a) Define the term **retailing**.

.... **(1 mark)**

Guided

(b) Study **Figure 1**.

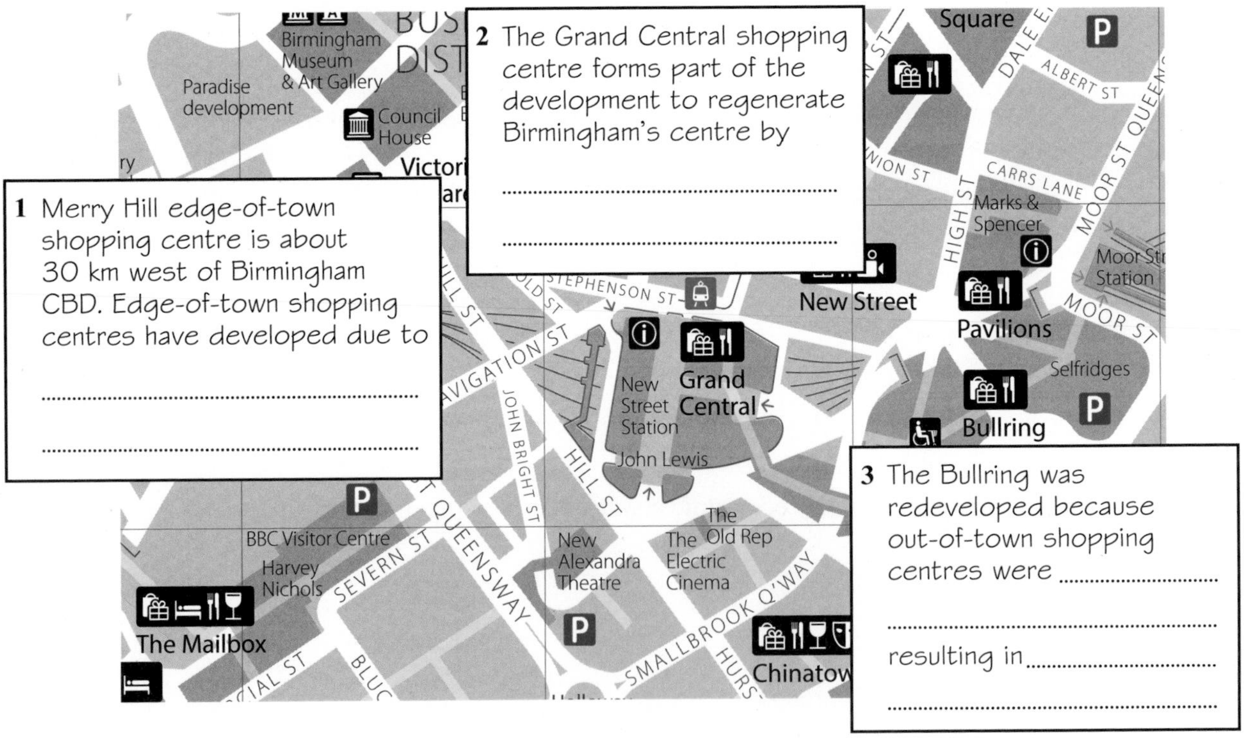

1 Merry Hill edge-of-town shopping centre is about 30 km west of Birmingham CBD. Edge-of-town shopping centres have developed due to ....

....

2 The Grand Central shopping centre forms part of the development to regenerate Birmingham's centre by

....

....

3 The Bullring was redeveloped because out-of-town shopping centres were ....

....

resulting in ....

....

Figure 1 Retail locations in Birmingham

Explain recent changes in retailing by completing **boxes 1–3** on **Figure 1**. **(3 marks)**

> This focuses on Birmingham, but you can apply your wider knowledge to complete the answers.

🌐 **Case study** **City living**

1   Study **Figure 1**.

> Birmingham City Council will have to lower the 2016 recycling target to just 30%.
> A recent report revealed that Birmingham only reached 29% household recycling rate for 2015, missing its 35% target.

Figure 1 Extract from a report about recycling in Birmingham

(a)   Define the term **recycling**.

.... **(1 mark)**

(b)   In 2015, Birmingham City Council recycled 26.6% of household waste. Calculate the percentage by which this failed to reach the 2015 target.

.... **(1 mark)**

Guided

(c)   Explain **one** reason why it is important for urban areas to carry out recycling.

Recycling reduces the need to send waste to landfill sites. This is important because

....

.... **(2 marks)**

2   Explain the differences between **affordable** and **energy-efficient** housing.

> Read questions carefully. Here you need to give **more than one** difference in your answer.

....

....

....

....

.... **(4 marks)**

3   Suggest **one** reason why sustainable transport can improve the quality of living in urban areas.

....

.... **(2 marks)**

Had a go ☐ Nearly there ☐ Nailed it! ☐

Case study Context and structure

You will have studied a named city in a developing or emerging country. As you work through the following pages, answer using your own case study city where possible.

1 Explain (a) the **site** and (b) the **situation** of a named city in a developing or an emerging country.

Named city in a developing or emerging country ....

(a) ....

....

(b) ....

.... **(4 marks)**

Guided

2 Study **Figure 1**.

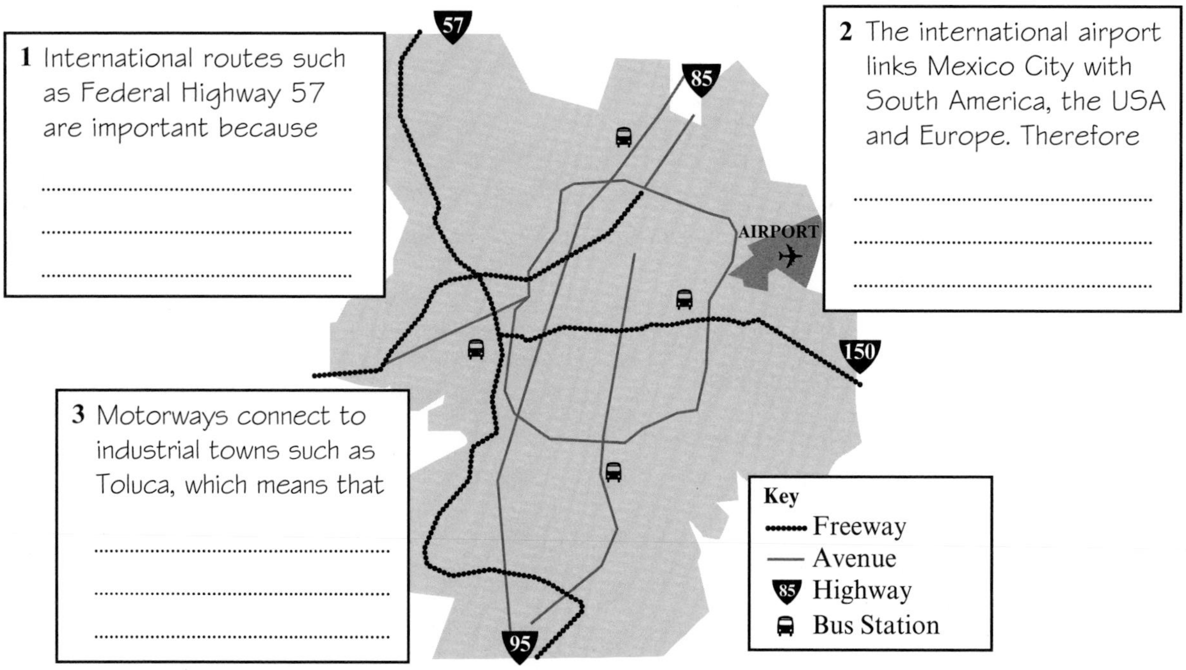
1 International routes such as Federal Highway 57 are important because
....
....
....
2 The international airport links Mexico City with South America, the USA and Europe. Therefore
....
....
....
3 Motorways connect to industrial towns such as Toluca, which means that
....
....
....
57
85
AIRPORT
150
95
Key
Freeway
Avenue
85 Highway
Bus Station

Figure 1 The connectivity of Mexico City

Explain the connectivity of urban areas such as Mexico City by completing **boxes 1–3** on **Figure 1.**

This focuses on Mexico City, but you can apply your wider knowledge to complete the answer.

(3 marks)

3 Explain the function of the CBD of a named city.

Remember to write about a city in a developing or emerging country.

Named city in a developing or emerging country ....

....

....

....

.... **(4 marks)**

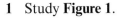

🌐 Case study A rapidly growing city

1 Study **Figure 1**.

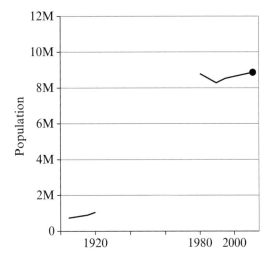

Figure 1 Population changes in Mexico City

(a) Plot the data in the table on to **Figure 1**.

Year	Population
1940	1.8 million
1960	9 million

> Remember to complete the line graph when you have plotted the data.

(2 marks)

(b) Explain why (i) natural increase and (ii) economic investment have resulted in the population growth of a named city.

Named city in developing or emerging country ....

(i) ....

....

(ii) ....

.... **(4 marks)**

2 Which **one** of the following gives the meaning of the term **national migration**?

☐ **A** People come to a country from another country

☐ **B** People move within the same country or region

☐ **C** People leave a country

☐ **D** People apply to become citizens   **(1 mark)**

Guided **3** Assess the impacts of migration on a named city.

Named city in developing or emerging country ....

One of the most evident impacts of migration is the rapid increase in the city's

population size. This causes demand for housing, which ....

....

.... **(8 marks)**

> Continue your answer on your own paper. You should aim to write approximately one side of A4.

Had a go ☐ Nearly there ☐ Nailed it! ☐

Case study Increasing inequalities

1 Explain **one** reason for the development of squatter settlements.

....

.... **(2 marks)**

Guided

2 Explain why rapid urbanisation might increase pollution in a named city.

Named city in developing or emerging country ....

A rapid increase in population and urbanisation leads to an increase in cars and other vehicles. Emissions from vehicles ....

.... **(4 marks)**

3 Study **Figure 1**.

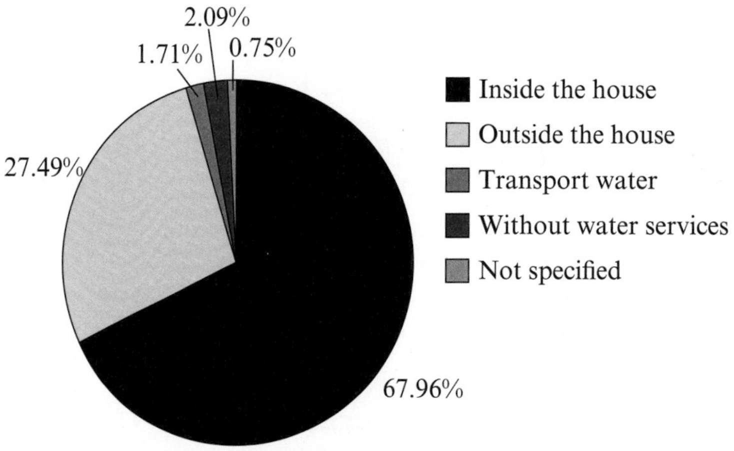

Figure 1 Domestic water supply, Mexico City

Explain why a large percentage of homes in cities in developing or emerging countries do not have internal water supplies.

Use data to support your answer.

....

....

....

.... **(4 marks)**

4 State the meaning of the term **under-employment**.

.... **(1 mark)**

5 Explain how the growth of a named city results in increasing inequality.

Named city in developing or emerging country ....

....

....

....

.... **(4 marks)**

🌐 Case study Solving city problems

1   Suggest **two** advantages of using a bottom-up approach to solve problems in a named city.

> You are only asked to talk about advantages.

Named city in developing or emerging country ....

....

....

....

.... **(4 marks)**

2   Explain how community-based projects are helping to solve problems in a named city in a developing or emerging country.

Named city in developing or emerging country ....

....

....

....

.... **(4 marks)**

Guided   3   Study **Figure 1**.

> • Mexico City's air has changed from being the world's cleanest to the dirtiest in 80 years
> • In the 1940s, the average visibility was 100 km; today it is about 1.5 km
> • Levels of pollutants like nitrogen dioxide (NO_2) regularly break international standards by two to three times

Figure 1 A slide from a presentation about urban air quality

Explain how the government of a named city in a developing or emerging country is trying to improve air quality.

> This focuses on Mexico City. Apply your wider knowledge to complete the answer.

Named city in developing or emerging country ....

Governments are introducing or expanding public transport systems to reduce air

pollution. For example, ...

....

.... **(4 marks)**

4   Suggest how a named city in a developing or emerging country is trying to reduce waste.

Named city in developing or emerging country ....

....

....

....

.... **(4 marks)**

Had a go ☐ Nearly there ☐ Nailed it! ☐

Defining development

1 Define the term **development**.

.... **(1 mark)**

2 Suggest **two** ways in which a low income country might develop.

....

....

....

.... **(4 marks)**

3 State **one** factor that contributes to increasing the human development of a country.

.... **(1 mark)**

4 (a) Define the term **water security**.

You should be able to define key development terms.

.... **(1 mark)**

Guided

(b) Study **Figure 1**.

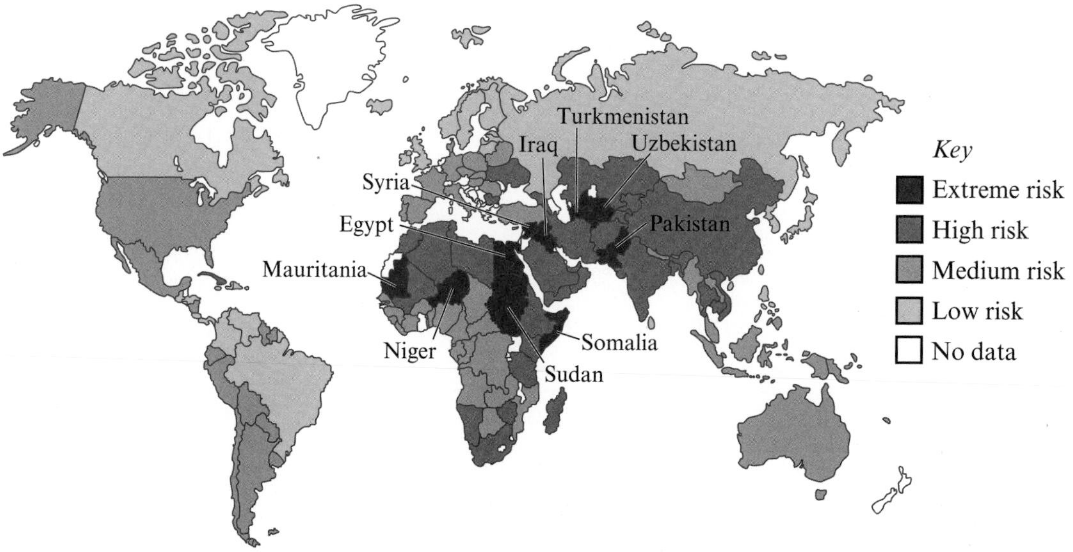

Figure 1 Global water security

Describe the distribution of countries at extreme risk of having poor water security.

Most of the countries are north of the Equator in a band that stretches from Africa

....

....

.... **(3 marks)**

5 Explain **one** reason why food security contributes to the human development of a country.

....

.... **(2 marks)**

Measuring development

1 Identify which **one** of the following factors is **not** a measure of development.

☐ **A** Indices of political corruption   ☐ **C** GDP

☐ **B** CBD   ☐ **D** HDI   **(1 mark)**

2 Study **Figure 1**.

Rank	Country	Index of political corruption	Rank	Country	Index of political corruption
1	Denmark	91	7	Switzerland	86
2	Finland	90	8	Singapore	85
3	Sweden	89	9	Canada	83
4	New Zealand	88	10	Germany	81
5	Netherlands	87	10	Luxembourg	81
5	Norway	87	10	United Kingdom	81

Figure 1 Indices of political corruption for some of the least corrupt countries

(a) Calculate the mode for the corruption perception index figures. Show your working.

(2 marks)

(b) Suggest how indices of political corruption might be used to measure development.

....

.... **(2 marks)**

3 Compare the use of the Gross National Product per capita and the Human Development Index to measure development.

> **Compare** means include similarities and differences.

....

....

....

.... **(4 marks)**

Guided **4** Explain how measures of inequality can be used as a measure of development.

Inequality measurements are often based on the distribution of income and economic

inequality among people living in a country or area. One way of measuring this

....

.... **(4 marks)**

69

Had a go ☐   Nearly there ☐   Nailed it! ☐

Patterns of development

1 (a) Which **one** of the following is the correct meaning of HDI?

    ☐ **A** A measure of development based on social and economic levels

    ☐ **B** A measure of development based on social and political levels

    ☐ **C** A measure of development based on GDP and economic levels

    ☐ **D** A measure of development based on social and political levels **(1 mark)**

(b) Study **Figure 1**.

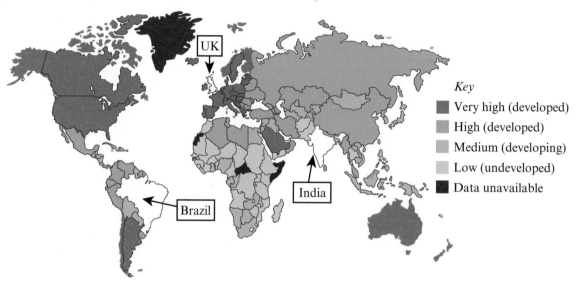

Key
- Very high (developed)
- High (developed)
- Medium (developing)
- Low (undeveloped)
- Data unavailable

Figure 1 A choropleth map to show the global distribution of HDI

Complete the choropleth map using the information in the table.

Country	Level of development
UK	Very high
Brazil	High
India	Developing

(3 marks)

Guided

(c) Describe the global distribution of HDI.

Most of the highly developed countries are in the northern hemisphere, especially

in northern America and Europe. Highly developed countries in the southern

hemisphere tend to be dispersed, ...

....

.... **(3 marks)**

2 Explain **two** reasons for spatial variations in levels of economic development in the UK.

> You need to know more than two reasons, but you only need to explain **two** of these in your answer.

....

....

....

.... **(4 marks)**

Uneven development

1 Define the term **quality of life**.

.... **(1 mark)**

> **Guided**

2 Suggest why uneven development has an impact on people's health.

People in developing countries often have very limited access to health care

.... **(2 marks)**

3 Study **Figure 1**.

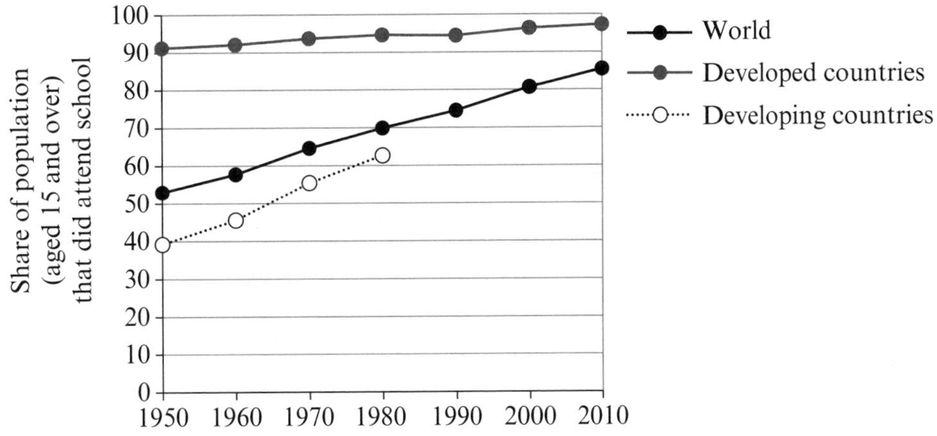

Figure 1 Percentage of the population over 15 years of age who attend school

(a) The data in the table show the percentage of the population over 15 years of age who attend school in developing countries. Using these data, complete the graph in **Figure 1**.

Year	%
1990	65
2000	72
2010	76

(3 marks)

(b) Suggest why there are differences in the percentage of the population over 15 years who attend school in developed and developing countries.

> Use data from the graph to show the difference, and then suggest reasons.

....

.... **(2 marks)**

4 Explain why uneven development has an impact on (i) food and (ii) water security.

(i) ....

....

(ii) ....

.... **(4 marks)**

Had a go ☐   Nearly there ☐   Nailed it! ☐

International strategies

1 Explain what is meant by the term **international aid**.

> Make sure you know the difference between **international aid** and **inter-governmental agreements**.

....

.... **(2 marks)**

2 Explain **two** ways international aid can help development.

....

....

....

.... **(4 marks)**

3 Study **Figure 1**.

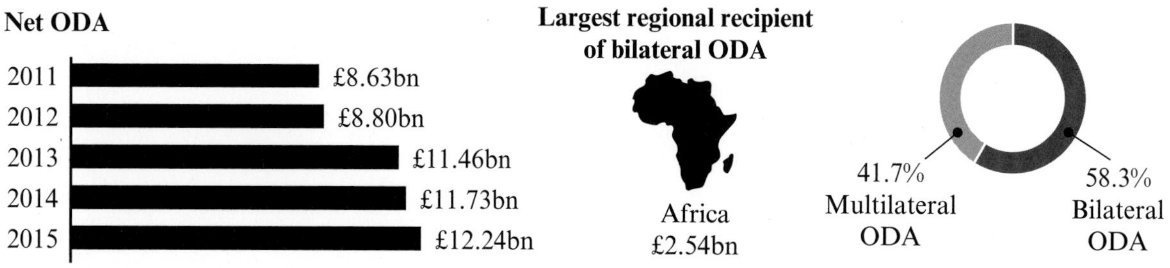

Net ODA

2011	£8.63bn
2012	£8.80bn
2013	£11.46bn
2014	£11.73bn
2015	£12.24bn

Largest regional recipient of bilateral ODA

Africa
£2.54bn

41.7%
Multilateral ODA

58.3%
Bilateral ODA

Figure 1 UK foreign aid 2015

> **Guided**

(a) Calculate the mean net ODA between 2011 and 2015. Show your working.

£8.63 bn + 8.80 bn + ... + ... + ... = ...

.... ÷ 5 = £............. bn **(2 marks)**

(b) Define the term **bilateral aid**.

.... **(1 mark)**

(c) Suggest why Africa is the largest recipient of UK bilateral ODA (aid).

....

.... **(2 marks)**

Top-down vs bottom-up

1 Explain how top-down development projects help to promote development.

....

....

....

.... **(4 marks)**

> **Guided** 2 Study **Figure 1**.

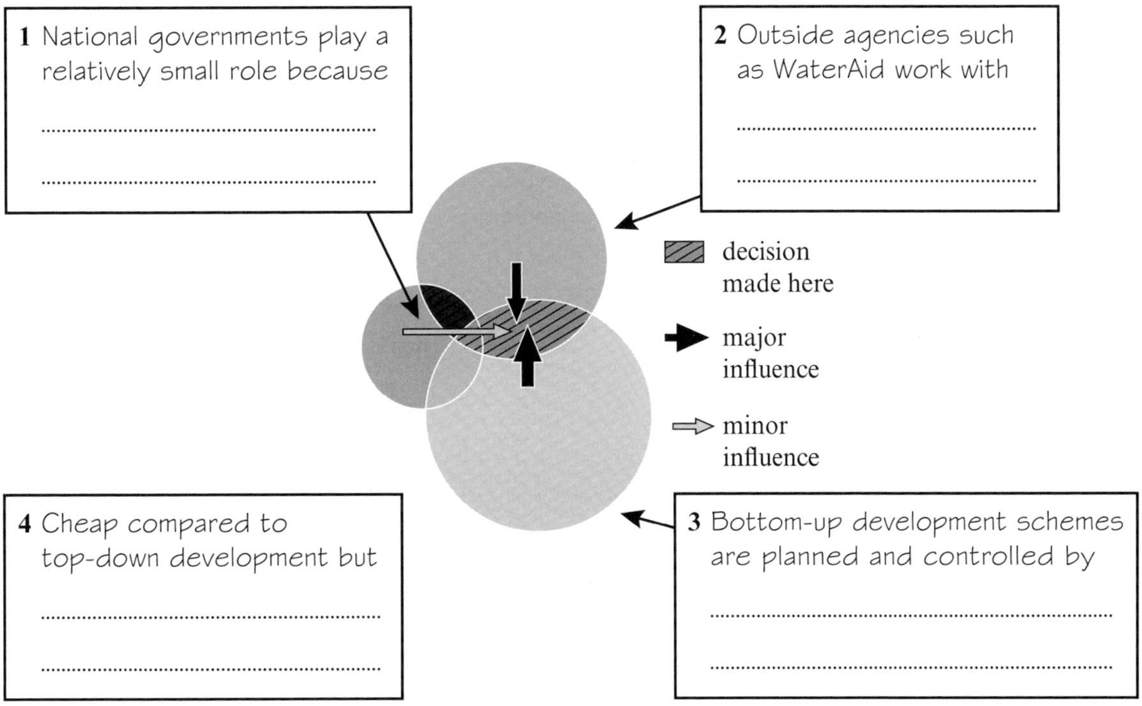

1 National governments play a relatively small role because

....

....

2 Outside agencies such as WaterAid work with

....

....

decision made here

major influence

minor influence

4 Cheap compared to top-down development but

....

....

3 Bottom-up development schemes are planned and controlled by

....

....

Figure 1 Bottom-up development

(a) Explain how bottom-up development projects help to promote development by completing **boxes 1–4** in **Figure 1**. **(3 marks)**

(b) Identify **one** disadvantage of bottom-up development projects.

☐ **A** Led by local people

☐ **B** Uses appropriate technology

☐ **C** Relatively inexpensive

☐ **D** Governments become over dependent on NGOs **(1 mark)**

> Read the question and options carefully – it's easy to make a mistake.

3 Explain the advantages and disadvantages of top-down led development projects.

....

....

....

.... **(4 marks)**

🌐 Case study **Location and context**

> You will have studied development in a named developing or emerging country. As you work through the following pages, answer using your own case study country where possible.

1   Describe the location of a developing or emerging country by constructing a labelled sketch map in the space below.

Named developing or emerging country ....

(3 marks)

> Guided

2   Explain the political context of your chosen country.

Named developing or emerging country ....

Emerging countries such as India are usually members of global groups including

the World Trade Organization and the United Nations. ....

....

....   **(4 marks)**

3   Identify **one** factor associated with the environmental context of your chosen country.

Named developing or emerging country ....

☐ **A** Population size   ☐ **C** Politics

☐ **B** Aid   ☐ **D** Climate   **(1 mark)**

4   Explain the social context of your chosen country.

> **Social** and **cultural** contexts are easily confused so make sure you know the differences.

Named developing or emerging country ....

....

....

....

....   **(4 marks)**

Case study Uneven development and change

1 Define the term **core**.

.... **(1 mark)**

2 Identify the meaning of the term **periphery.**

   ☐ **A** Regions with high levels of development

   ☐ **B** Regions with both high and low development

   ☐ **C** Regions in which uneven development occurs

   ☐ **D** Regions with lower levels of development   **(1 mark)**

⟩ **Guided** ⟩   **3** Study **Figure 1**.

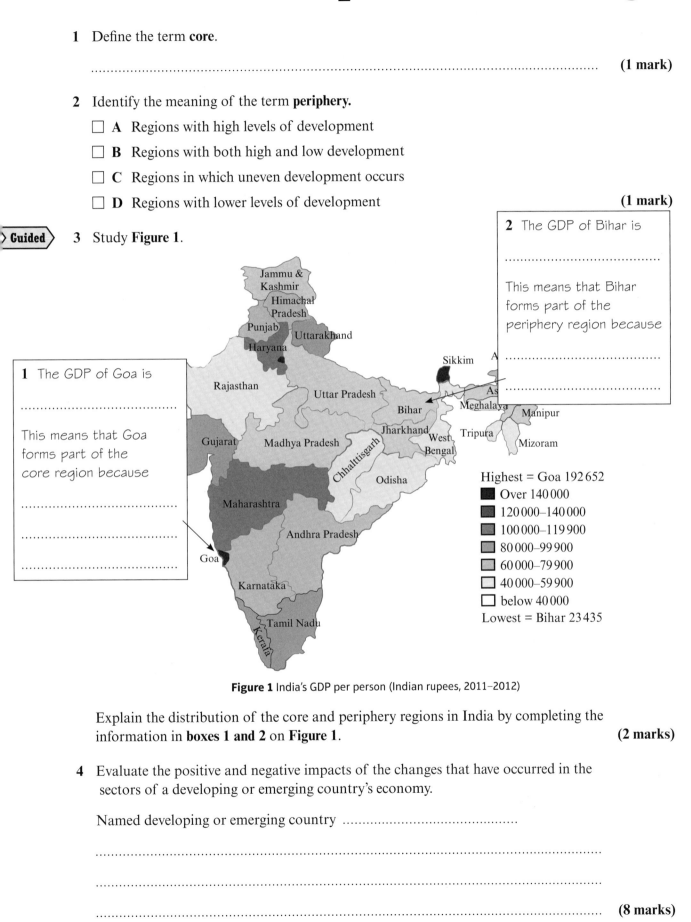

2 The GDP of Bihar is

....

This means that Bihar forms part of the periphery region because

....

....

1 The GDP of Goa is

....

This means that Goa forms part of the core region because

....

....

....

Highest = Goa 192 652
- ■ Over 140 000
- ■ 120 000–140 000
- ■ 100 000–119 900
- ▨ 80 000–99 900
- ▨ 60 000–79 900
- ☐ 40 000–59 900
- ☐ below 40 000

Lowest = Bihar 23 435

Figure 1 India's GDP per person (Indian rupees, 2011–2012)

Explain the distribution of the core and periphery regions in India by completing the information in **boxes 1 and 2** on **Figure 1**.   **(2 marks)**

4 Evaluate the positive and negative impacts of the changes that have occurred in the sectors of a developing or emerging country's economy.

Named developing or emerging country ....

....

....

.... **(8 marks)**

┌──┐
│ Continue your answer on your own paper. You should aim to write approximately one side of A4. │
└──┘

Had a go ☐   Nearly there ☐   Nailed it! ☐

Case study Trade, aid and investment

> **Guided**

1 Explain what is meant by the term **international trade**.

International trade is exchanging goods, services ....

.... **(2 marks)**

2 Study **Figure 1**.

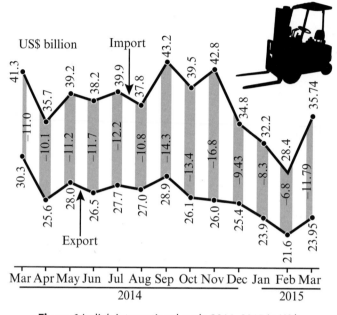

Figure 1 India's international trade 2014–2015 in US$

(a) Compare the graphs for India's exports and imports between March 2014 and March 2015.

> Remember to include general patterns, figures and anomalies.

....

....

.... **(3 marks)**

(b) Explain **one** characteristic of international trade for a developing or emerging country.

Named developing or emerging country ....

....

.... **(2 marks)**

3 Suggest reasons for variations in the amount of aid received by a developing or emerging country.

Named developing or emerging country ....

....

....

....

.... **(4 marks)**

🌐 Case study **Changing population**

1 (a) Identify **one** characteristic shown in a population pyramid.

☐ **A** Population distribution   ☐ **C** Population structure

☐ **B** Population density   ☐ **D** Population size   **(1 mark)**

(b) Study **Figure 1**.

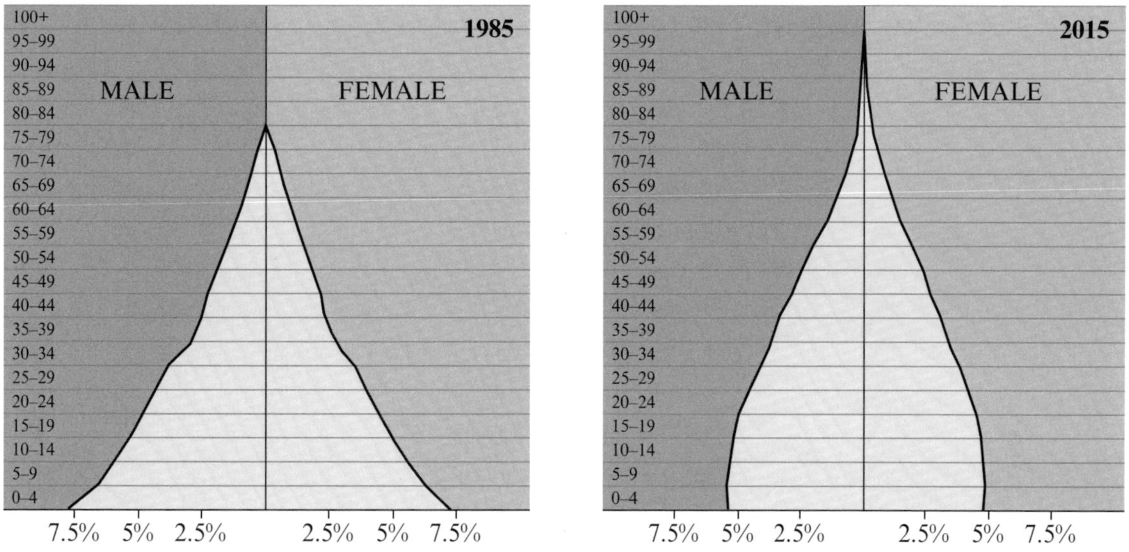

Figure 1 Population pyramids for India, 1985 and 2015

> **Guided**

Compare the population pyramids for 1985 and 2015.

In 1985, the population pyramid was triangular in shape

| **Compare** means look for similarities and differences. |

....

.... **(3 marks)**

2 Explain why life expectancy in a named developing or emerging country has changed over the last 30 years.

Named developing or emerging country ....

....

....

....

.... **(4 marks)**

3 Suggest the impacts of (a) a growing middle class and (b) improved education on a developing or emerging country.

Named developing or emerging country ....

(a) ....

....

(b) ....

.... **(4 marks)**

⊕ Case study Geopolitics and technology

Guided   1   Explain the meaning of the term **geopolitics**.

Geopolitics is the impact of a country's human and physical geography ....

.... **(2 marks)**

2   Suggest how the development of a developing or emerging country is affected by geopolitics.

....

....

....

.... **(4 marks)**

3   Study **Figure 1**.

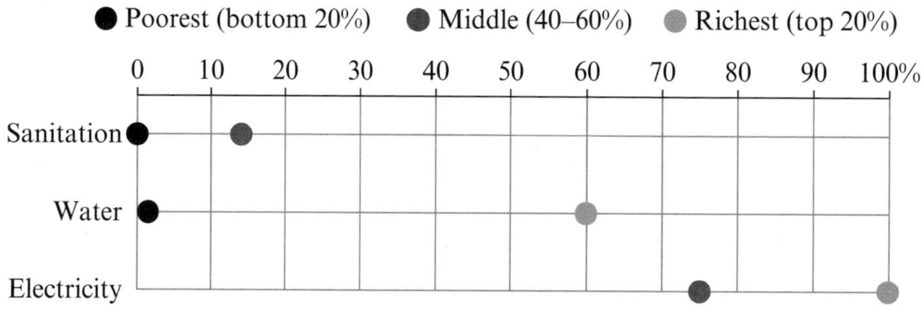

● Poorest (bottom 20%)   ● Middle (40–60%)   ● Richest (top 20%)

Figure 1 Access to basic services in India 2015

(a)   Plot the information in the table on to **Figure 1**.

Sanitation	Top 20%	84%
Water	Middle 40–60%	15%
Electricity	Bottom 20%	16%

(3 marks)

(b)   Which **one** of the following groups of children would be most likely to live in the core region of India?

☐ **A** Top 20% and bottom 20%   ☐ **C** Middle 40–60% and bottom 20%

☐ **B** Middle 40–60%   ☐ **D** Top 20%   **(1 mark)**

4   Explain how improving connectivity can support development in different parts of a named developing or emerging country.

Named developing or emerging country ....

....

....

....

.... **(4 marks)**

Case study **Impact of rapid development**

1   Explain **two** negative environmental impacts of rapid development for a developing or emerging country.

> You need to know positive and negative impacts, but only write about the negative ones in this answer.

Named developing or emerging country ....

....

....

....

.... **(4 marks)**

2   State **one** positive economic impact of rapid development.

.... **(1 mark)**

Guided   **3**   Study **Figure 1**.

> **2** These areas are often polluted and have poor air quality. This means that
> ....
> ....

> **1** Rapid development means that people move to urban areas. This causes
> ....
> ....

> **3** There are no or very few services or roads for medical treatment access. Diseases such as
> ....
> ....

Figure 1 A shanty town in Jammu, India

Explain the negative social impacts of rapid development by completing **boxes 1–3** on **Figure 1**.   **(3 marks)**

4   Examine how a developing or emerging country's government and people are managing the impacts of rapid development in order to improve the quality of life.

Named developing or emerging country ....

....

....

....

.... **(8 marks)**

> Continue your answer on your own paper. You should aim to write approximately one side of A4.

Had a go ☐ Nearly there ☐ Nailed it! ☐

The world's natural resources

1 Define the term **natural resource**.

.... **(1 mark)**

2 Explain the difference between biotic and abiotic resources.

....

.... **(2 marks)**

3 Identify which **two** of the following are characteristics of renewable resources.

☐ **A** Potentially inexhaustible

☐ **B** Take million of years to form

☐ **C** Examples include natural gas

☐ **D** Can be naturally replenished **(2 marks)**

4 Study **Figure 1**.

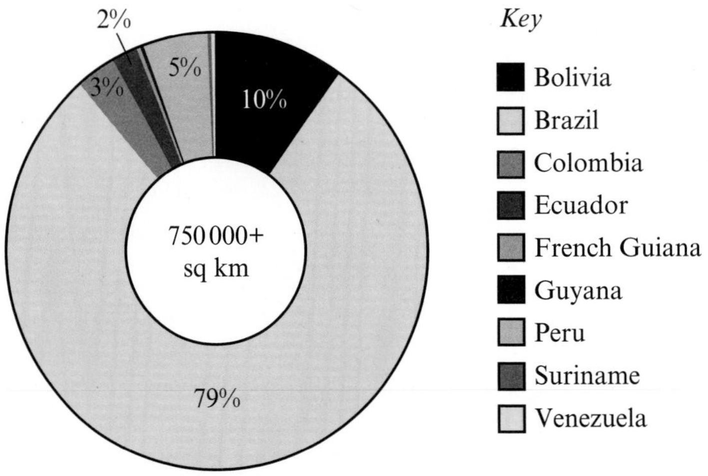
2%
5%
3%
10%
750 000+
sq km
79%
Key
Bolivia
Brazil
Colombia
Ecuador
French Guiana
Guyana
Peru
Suriname
Venezuela

Figure 1 Accumulated loss of Amazon rainforest 1978–2014

(a) Calculate the accumulated loss of rainforest in Bolivia in square kilometres (km^2). Show your calculations.

Remember to look carefully at the key.

(2 marks)

Guided

(b) Explain why natural environments such as rainforests are exploited by people.

Natural environments are exploited to obtain resources, for example the rapid deforestation of tropical forests produces high-quality timber, which is a valuable export. Other reasons for exploitation are ....

....

.... **(4 marks)**

Variety and distribution

1 Study **Figure 1**.

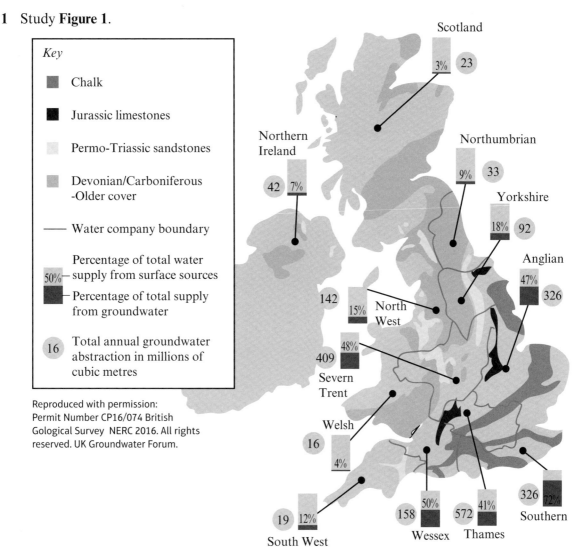

Figure 1 Located divided bar graphs to show the origins of water supply in the UK

(a) State which region obtains the highest percentage of its water from groundwater supplies.

> You need to look carefully at the key and the divided bar graphs.

.... **(1 mark)**

> Guided

(b) Suggest why groundwater abstraction is highest in the Thames region of the UK.

The Thames region contains much of Greater London and the Thames Valley,

including large towns such as Reading. Therefore the population density

.... **(2 marks)**

2 Explain **two** factors that influence the global distribution of agriculture.

....

....

....

.... **(4 marks)**

Global usage and consumption

1 Study **Figure 1**.

Region	1984–1986	1997–1999	2015	2030
World	2655	2803	2940	3050
Developing/emerging countries	2450	2681	2850	2980
Developed countries	3206	3380	3440	3500

Figure 1 Food consumption (Kcal per person per day)

(a) Calculate the projected difference in food consumption between developing/emerging countries and developed countries in 2030.

.... **(1 mark)**

(b) Explain **one** reason for this difference.

....

.... **(2 marks)**

2 (a) Identify **one** reason for the uneven global use of fuels.

.... **(1 mark)**

(b) Study **Figure 2**.

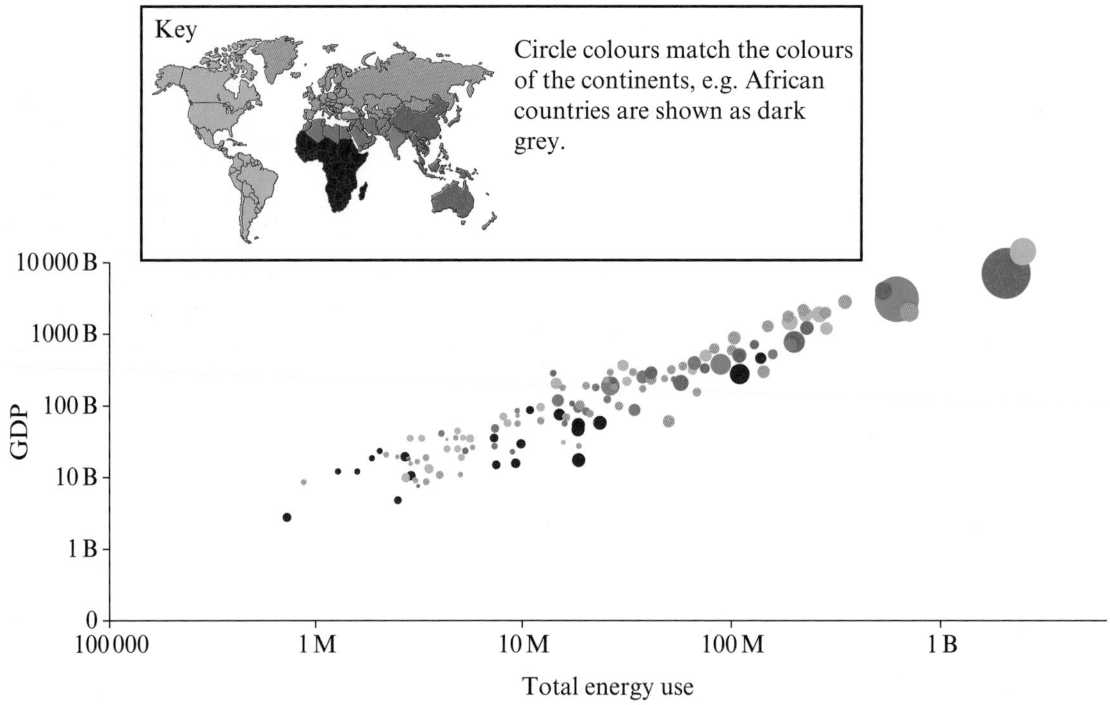

Key

Circle colours match the colours of the continents, e.g. African countries are shown as dark grey.

Figure 2 Energy per country use plotted as proportional circles against GDP (Gross Domestic Product)

(i) Using **Figure 2**, identify the continent that contains the country that uses the least energy.

Use the colour of the circles and the world map.

.... **(1 mark)**

Guided

(ii) Suggest why energy use tends to be low in the continent you identified in (i).

This continent includes a large number of countries that are either emerging

or developing. Therefore ....

.... **(2 marks)**

Production and development

Only revise pages 83–88 if you studied Energy resource management.

1 Explain the difference between renewable and non-renewable energy resources.

> You do not need to give examples of renewable and non-renewable energy resources, but these might help your answer.

....

.... **(2 marks)**

2 Explain the disadvantages of the production and development of **one** non-renewable energy resource.

....

....

....

.... **(4 marks)**

Guided

3 Study **Figure 1**.

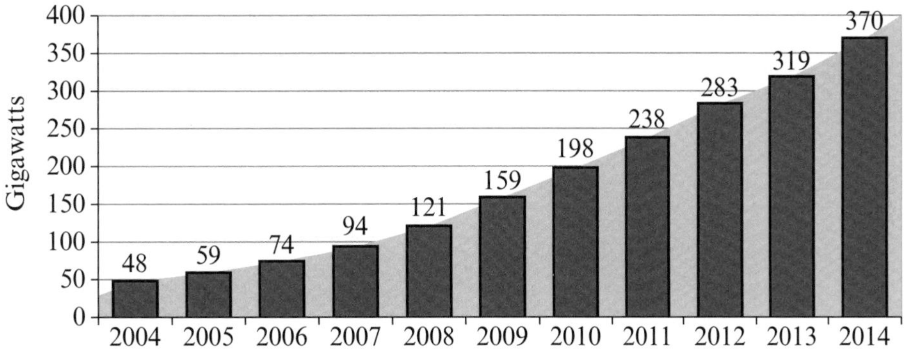

Figure 1 Potential global wind power production, 2004–2014

Describe the changes in potential global wind power production between 2004 and 2014.

There has been a steady increase in potential global wind power production

between 2004 and 2014. ....

....

.... **(3 marks)**

4 State **one** advantage of using coal as an energy resource.

.... **(1 mark)**

5 Explain **one** disadvantage of using a non-renewable energy resource.

....

.... **(2 marks)**

UK and global energy mix

1 Explain the term **energy mix**.

....

.... **(2 marks)**

2 Explain why the composition of the UK's energy mix has changed since 1970.

....

....

....

.... **(4 marks)**

3 Study **Figure 1**.

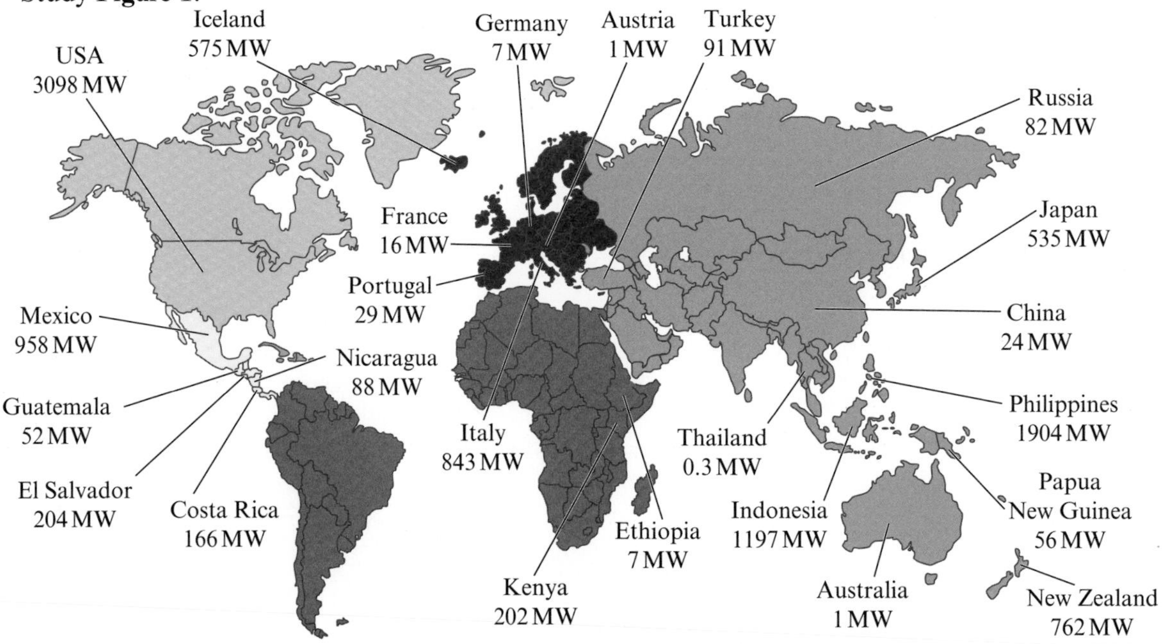

Figure 1 Global geothermal power, 2010

  (a) State the country which produced the most geothermal power in 2010.

.... **(1 mark)**

Guided

  (b) Explain why global variations in the energy mix depend on availability.
     Include information from **Figure 1** in your answer.

Countries with large resources of one type of energy do not have a true energy

mix but depend on that resource as it is usually the cheapest. For example,

New Zealand and Iceland rely on ....

.... **(2 marks)**

4 Global population growth is predicted to exceed energy production by 2075.
Explain **one** other reason for changes in global energy demand.

....

.... **(2 marks)**

Impacts of non-renewable energy resources

1   Study **Figure 1**.

(a)   State which energy type uses uranium as a fuel source.

....

(1 mark)

> Make sure you are familiar with a range of graphs – see page 126.

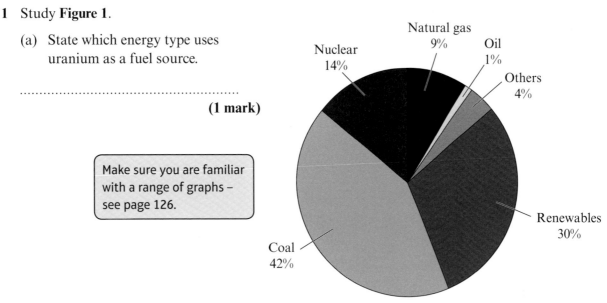

Nuclear 14%
Natural gas 9%
Oil 1%
Others 4%
Renewables 30%
Coal 42%

Figure 1 Energy production mix in Germany 2015

> **Guided**

(b)   Calculate the percentage of energy obtained from fossil fuels. Show your calculations.

(Coal) 42% + (oil) .... + ...   **(1 mark)**

(c)   Suggest the negative effects that developing fossil fuels can have on the environment.

> Make sure you know the positive effects as well.

....

....

....

....   **(4 marks)**

2   State **one** impact on people of using uranium as a fuel source.

....   **(1 mark)**

3   Explain how technology such as fracking might resolve energy resource shortages.

....

....

....   **(3 marks)**

Impacts of renewable energy resources

1 (a) Wind power is a renewable energy resource.
Name **one** other renewable energy resource.

> For this type of question you only need to answer in two or three words.

.... **(1 mark)**

> **Guided**

(b) Identify **two** statements that describe the positive impacts of wind power.

☐ **A** Can cause noise and visual pollution

☒ **B** Reduces CO_2 emissions

☐ **C** Can generate enough energy for thousands of homes

☐ **D** May impact on the migration patterns of birds **(2 marks)**

2 Suggest **one** disadvantage of developing of HEP (hydro-electric power).

....

.... **(2 marks)**

3 State **one** advantage of developing wind power.

.... **(1 mark)**

4 Study **Figure 1**.

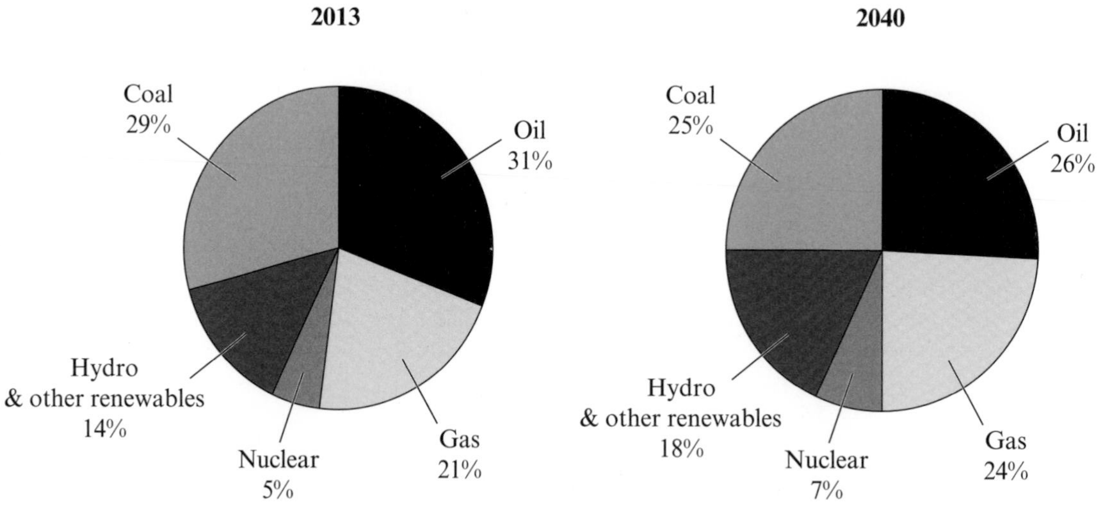

2013   **2040**

Figure 1 Actual and projected global energy demand

(a) Calculate the difference between renewable energy demand in 2013 and the projected demand in 2040.

.... **(1 mark)**

(b) Suggest why there is a projected change in demand.

....

.... **(2 marks)**

Meeting energy demands

Guided   **1**   Suggest why it is important to manage energy resources.

Scientists suggest relying on fossil fuels could have irreversible impacts caused by

climate change such as sea level rises which ....

....

....

.... **(4 marks)**

2   Study **Figure 1**.

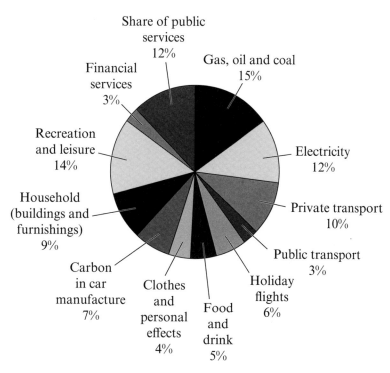

Figure 1 A typical UK person's carbon footprint, 2015

From **Figure 1**, identify the source of the highest carbon emissions.

☐ **A**   Recreation and leisure   ☐ **B**   Gas, oil and coal

☐ **C**   Electricity   ☐ **D**   Holiday flights   **(1 mark)**

3   Examine how attitudes to the exploitation and consumption of energy resources vary between different stakeholders.

> Remember to include a range of views, such as those of individuals and governments.

....

....

.... **(8 marks)**

> Continue your answer on your own paper. You should aim to write approximately one side of A4.

87

🌐 Located example **China and Germany**

1   Explain why it is important for a named developed country to manage their energy resources.

Named developed country ....

....

....

....

.... **(4 marks)**

> **Guided**

2   Study **Figure 1**.

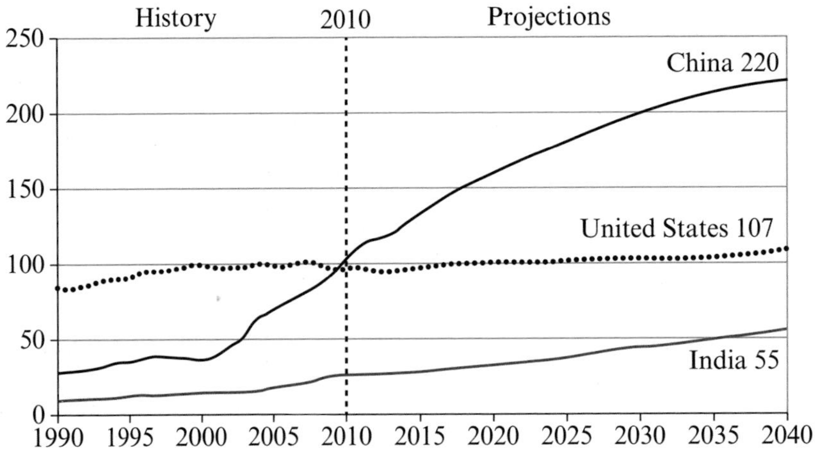

Figure 1 Actual and predicted energy consumption in quadrillion British thermal units

Compare the actual and predicted energy consumption for the United States and China.

Both the actual and predicted energy consumption for China increase much more

rapidly than the consumption of the USA. ....

....

.... **(3 marks)**

3   Assess how successful countries at different levels of development have been in managing their energy resources in a sustainable way.

> Make sure you write about both the emerging/developing country you have studied and a developed country.

....

....

....

.... **(8 marks)**

> Continue your answer on your own paper. You should expect to write about one side of A4.

Global distribution of water

Only revise pages 89-96 if you studied Water resource management.

1 Define the term **fresh water**.

.... **(1 mark)**

2 Study **Figure 1**.

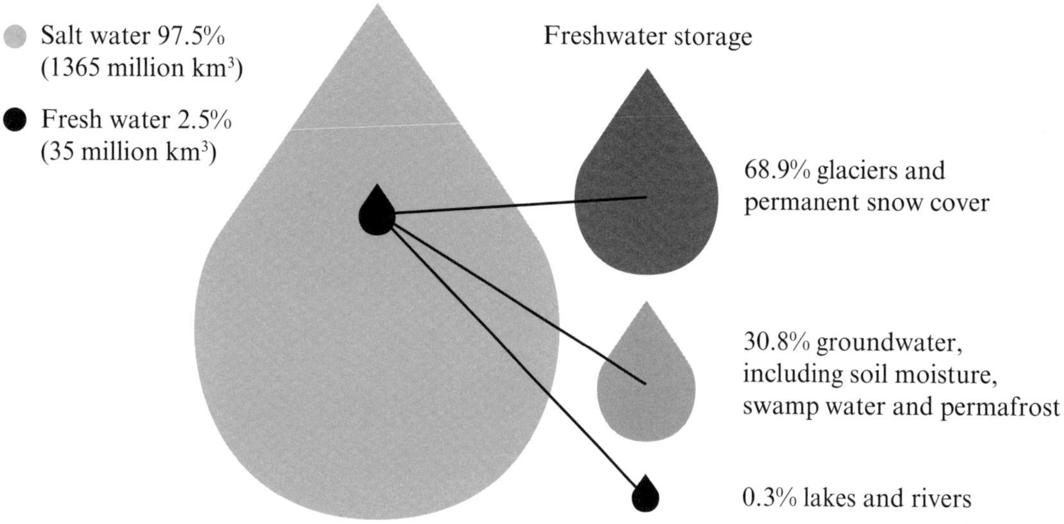

● Salt water 97.5%
   (1365 million km³)

● Fresh water 2.5%
   (35 million km³)

Freshwater storage

68.9% glaciers and
permanent snow cover

30.8% groundwater,
including soil moisture,
swamp water and permafrost

0.3% lakes and rivers

Figure 1 Estimated amounts of global saline and fresh water by percentage

(a) Identify which of the stores shown in **Figure 1** contains the highest percentage
of fresh water.

.... **(1 mark)**

> Guided

(b) Explain why only about 1 per cent of fresh water is available for use.
Use data from **Figure 1** in your answer.

Most of the fresh water is in stores that make it impossible to use. For example,

nearly 70 per cent ....

.... **(2 marks)**

3 Define the term **water surplus**.

.... **(1 mark)**

4 Suggest reasons why one named global region has
a water deficit.

Named global region ....

You need to know about the
availability of water on a
global, national and local scale.

....

....

....

.... **(4 marks)**

89

Had a go ☐ Nearly there ☐ Nailed it! ☐

Changing water use

1 Which **one** of the following describes how water demand has changed over the last 50 years?

☐ **A** Increased

☐ **B** Stayed the same

☐ **C** Increased slightly

☐ **D** Decreased

(1 mark)

Guided

2 Suggest reasons why water demand in Asia has increased since 2010.

Asia includes developing countries such as India, where rapidly increasing populations are increasing food demands. More water is needed for

.... **(2 marks)**

3 Study **Figure 1**.

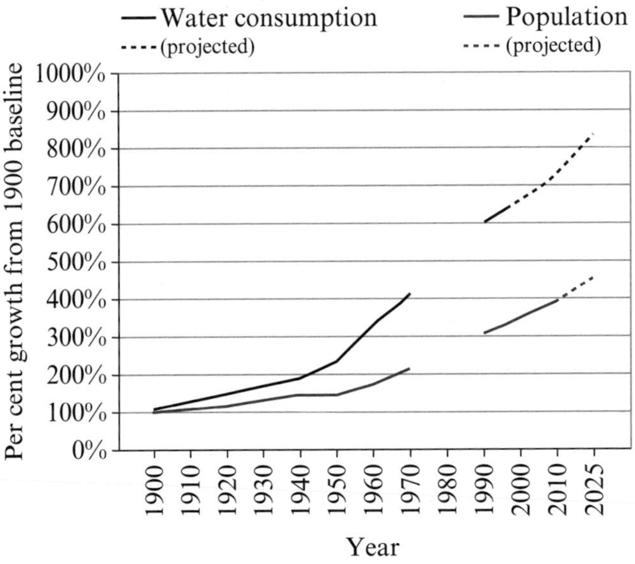

Figure 1 Actual and predicted global water demand and population growth. 1900 figures represent 100%

(a) Plot the data in the table to complete the population graph in **Figure 1**. **(3 marks)**

Year	Population (%)
1970	200
1980	250
1990	300

Use a pencil for plotting, then you can rub out any mistakes.

(b) Suggest reasons why global water demand has increased more rapidly than population growth. Include information from **Figure 1** in your answer.

....

....

....

.... **(4 marks)**

Water consumption differences

1   Explain how the mean water consumption for a country is calculated.

....

.... **(2 marks)**

Guided   2   Study **Figure 1**.

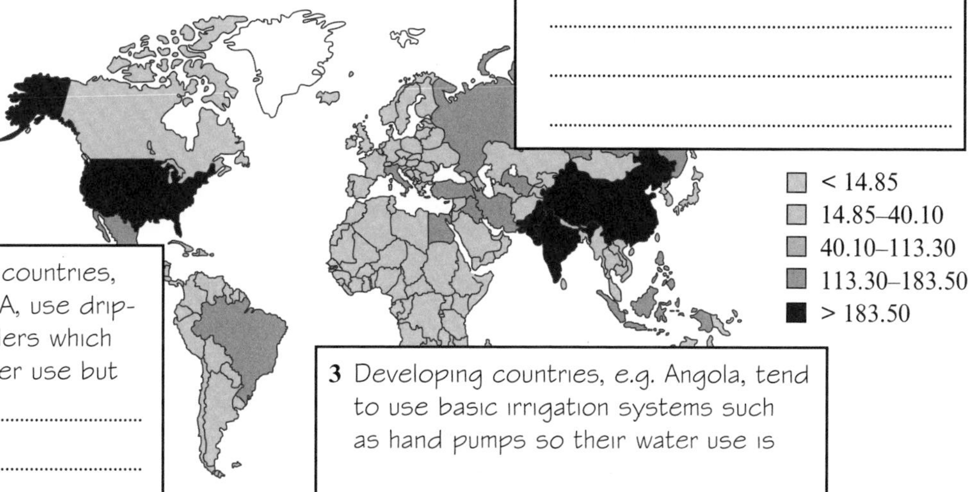

2 Emerging countries, e.g. India, rely on domestic agriculture for food production which means that

....

....

....

☐ < 14.85
☐ 14.85–40.10
☐ 40.10–113.30
☐ 113.30–183.50
■ > 183.50

1 Developed countries, e.g. the USA, use drip-feed sprinklers which reduce water use but

....

....

....

....

3 Developing countries, e.g. Angola, tend to use basic irrigation systems such as hand pumps so their water use is

....

....

....

Figure 1 Annual freshwater usage 2014 (billion cubic metres)

Explain why there are differences between the amount of water used for agriculture in developed countries and the amount used in emerging and developing countries by completing **boxes 1–3** on **Figure 1**.   **(3 marks)**

3   Identify **two** reasons for the relatively low use of water for domestic purposes in emerging/developing countries.

☐ **A** People bathe in streams and rivers

☐ **B** Domestic appliances are very efficient

☐ **C** Metered water is expensive

☐ **D** People's homes lack piped water   **(2 marks)**

4   Compare the amounts of water used by industry in developed and emerging or developing countries.

> You don't have to write about named countries unless the question specifies this.

....

....

....

.... **(4 marks)**

Water supply problems: UK

1 Explain why the imbalance in supply of rainfall in the UK may cause water supply problems.

Higher rainfall tends to occur in areas of lower population density, such as the

Highlands of Scotland and north Wales, so these areas do not have a water deficit

or a supply problem. However, ...

....

.... **(4 marks)**

2 Study **Figure 1**.

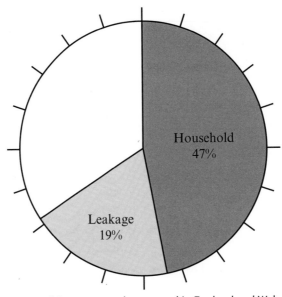

Figure 1 How public water supply was used in England and Wales (2011)

(a) Use the data in the table below to complete **Figure 1**.
     Label your completed pie chart.   **(3 marks)**

Services	10%
Manufacturing	4%
Other	20%

(b) Suggest **one** reason why nearly 20 per cent of the water supply was lost because of leakage.

....

.... **(2 marks)**

3 (a) Explain **one** cause of the seasonal imbalance of rainfall supply in the UK.

....

.... **(2 marks)**

(b) Suggest **one** problem that the seasonal imbalance of rainfall might cause for water supply companies.

.... **(1 mark)**

Water supply problems: emerging or developing countries

1 (a) Explain what is meant by the term **untreated water**.

....

.... **(2 marks)**

Guided

(b) Suggest why access to only untreated water is a problem in emerging or developing countries.

Without safe water it is very difficult for people to lead healthy and productive lives.

If people are not productive, food production is low and this means that

....

.... **(4 marks)**

(c) Study **Figure 1**.

Figure 1 Percentage of the global population without access to safe drinking water

Key:
- ☐ 0–6%
- ☐ 7–17%
- ☐ 18–32%
- ☐ 33–48%
- ☐ 49–63%
- ■ 64–87%
- ☐ No data

Identify the continent with the highest percentage of people without access to safe drinking water.

> You should know key global locations so that you can interpret maps like this.

☐ **A** South America 　　 ☐ **C** Africa

☐ **B** Europe 　　 ☐ **D** North America 　　 **(1 mark)**

2 (a) Name **one** area in an emerging or developing country that has a low annual rainfall.

.... **(1 mark)**

(b) Suggest why a low annual rainfall may cause water supply problems in emerging and developing countries.

....

.... **(2 marks)**

Attitudes and technology

1 (a) State the name of the process that removes salt from seawater.

.... **(1 mark)**

▷ **Guided** ▷ (b) Explain **one** advantage and **one** disadvantage of using this process to obtain fresh water.

One advantage of this method is that it reduces the demand for groundwater

extraction, which means that there is less impact on ecosystems ....

....

.... **(4 marks)**

2 Study **Figure 1**.

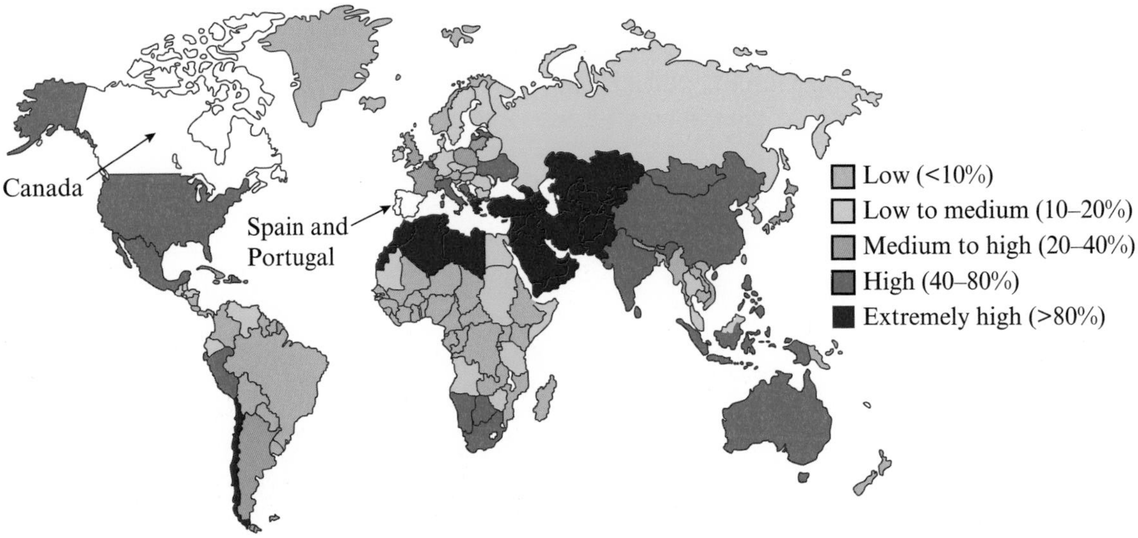

Canada

Spain and Portugal

Low (<10%)
Low to medium (10–20%)
Medium to high (20–40%)
High (40–80%)
Extremely high (>80%)

Figure 1 Projected water stress by country (2040)

(a) Use the information in the table to show the projected water stress levels for (i) Canada and (ii) Spain and Portugal.

> Look at the key for the correct colours to use.

Country	Water withdrawal: supply
Canada	Low to medium
Spain and Portugal	High

(2 marks)

(b) Describe the distribution of the areas predicted to be under severe water stress in 2040.

....

.... **(2 marks)**

3 Suggest reasons why attitudes to the exploitation and consumption of water resources vary between stakeholders.

....

....

....

.... **(4 marks)**

Managing water

1 Identify which **one** of the following gives the definition of water stress.

☐ **A** Water abstraction is the same as the amount available

☐ **B** Water abstraction is less than the amount available

☐ **C** Water abstraction is higher than the amount available

☐ **D** Water abstraction is difficult because of poor technology   **(1 mark)**

2 Define the term **sustainable management**.

> This is a very important term. Make sure you learn what it means!

....

.... **(1 mark)**

Guided

3 Suggest reasons why water resources need sustainable management.

Water is essential for life, and as global populations increase water resources are

increasingly under stress. Therefore ....

....

.... **(4 marks)**

4 Study **Figure 1**.

When full, Lake Mead is 373 metres high. In 2016 it was 327 metres high.

Lake Mead supplies 90 per cent of the water used by Las Vegas.

The 'bath tub 'ring shows that Lake Mead is only half full.

People who organise boat trips and fish depend on the lake for their income.

Figure 1 Information about Lake Mead which supplies water to Las Vegas, Nevada, USA

(a) Calculate the percentage change in the height of Lake Mead from being full and the level in 2016.

.... **(1 mark)**

(b) Suggest reasons why individuals might have different views about managing water resources such as Lake Mead in a sustainable way.

....

....

....

.... **(4 marks)**

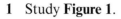 UK and China

1   Study **Figure 1**.

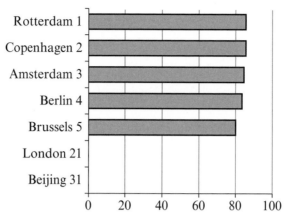

Water management index score

Figure 1 Sustainable cities rankings based on water management

(a) Plot the data in the table to show the index
rating for London (UK) and Beijing (China).

London	65
Beijing	61

(2 marks)

Guided

(b) Suggest reasons why cities in developed countries usually have a higher index for
sustainable water management than cities in emerging or developing countries.

Cities in developed countries have more finance available because ....

....

.... **(3 marks)**

2   Explain **two** ways in which a named developed country is trying to manage water
resources in a sustainable way.

Named developed country ....

....

....

....

.... **(4 marks)**

3   Assess the effectiveness of an emerging or developing country in
managing water resources in a sustainable way.

Named emerging or developing country ....

Think about why some
methods have worked
better than others.

....

....

....

....

.... **(8 marks)**

Continue your answer on your own paper. You should aim to write approximately one side of A4.

Paper 2

In this question, four additional marks will be awarded for your spelling, punctuation and grammar and for your use of specialist terminology.

Assess the importance of water supply problems for countries at different levels of development.

For this question you need to consider both the **developed** and the **emerging or developing countries** you have studied. You need to include a range of factors, including imbalances in water supply, ageing infrastructure, untreated water and low annual income. Use the information to decide and explain which of the factors are more important.

The content for Assessment Objective 2, which is worth 4 marks, might include the following explanations for water supply problems.

Developed countries (e.g. the UK)

- Most rainfall occurs in the northern and western regions where population density is low. Areas such as South East England, with a very high population density, have low rainfall totals.
- There are seasonal imbalances in rainfall and water demand – rainfall is low in the summer in the east of the UK when demand is high.
- Much of the infrastructure is over a hundred years old, which causes leaking pipes.

Developing or emerging countries (e.g. China)

- The lack of water treatment plants and rapid urbanisation means that many people do not have access to safe drinking water.
- Water courses are polluted as these act as sewage systems and rubbish dumps.
- Some areas such as the Sahel region of Africa have a low and unpredictable rainfall pattern.

For Assessment Objective 3, which is worth 4 marks, you would need to:

- **weigh up** the relative importance of both the social and economic factors causing the deforestation of rainforests
- **make a judgement** based on evidence that includes facts and figures.

For SPGST, which is worth 4 marks, you need to:

- **spell** (especially geographical terms) and punctuate accurately
- make sure you follow the rules of **grammar**
- use a wide range of **geographical terms**.

(8 marks plus 4 marks for SPGST)

Continue your answer on your own paper. You should aim to write about a side of A4.

Formulating enquiry questions

This section is about Coastal fieldwork. For Rivers fieldwork, see page 101.

Guided   **1**   Identify the stages in the enquiry process by completing boxes 2, 4 and 6 in the diagram below.

(3 marks)

| 1 Develop a question | → | 2 .... | → | 3 Process and present your data | → | 4 .... | → | 5 Make conclusions | → | 6 .... |

2   Study **Figure 1**.

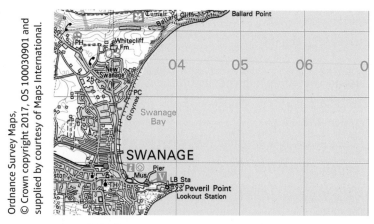

Ordnance Survey Maps,
© Crown copyright 2017, OS 100030901 and
supplied by courtesy of Maps International.

Figure 1 Ordnance Survey map 1:50 000 of Swanage Bay, Dorset

Some students intend to carry out fieldwork at Swanage Bay, shown in Figure 1.

(a)   Suggest an enquiry question based on coastal processes that they might investigate.

.... **(1 mark)**

(b)   Suggest **one** reason why Swanage Bay is a suitable location to investigate your chosen enquiry question.

      Look carefully at the map.

....

.... **(3 marks)**

3   (a)   State **one** theory or concept that might be investigated as part of this fieldwork.

.... **(1 mark)**

     (b)   Explain why the theory or concept named in (a) is relevant to this investigation.

....

....

....

.... **(4 marks)**

4   Study **Figure 1**. Name **one** man-made feature that might affect coastal process in the study area.

.... **(1 mark)**

Methods and secondary data

1 (a) Identify which **one** of the following gives a definition of **secondary data**.

☐ **A** Data collected using sampling

☐ **B** Data collected very quickly

☐ **C** Data collected by someone else

☐ **D** Data collected directly during fieldwork   **(1 mark)**

(b) Study **Figure 1**.

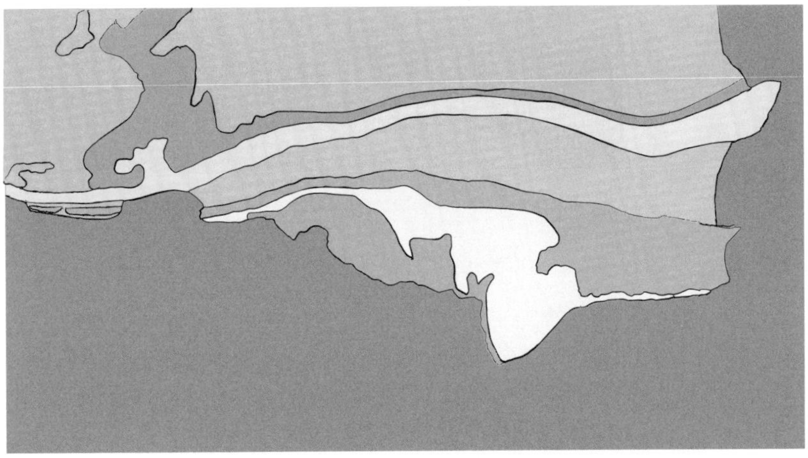

Figure 1 Geological map of the Swanage area

Explain how the information
shown in **Figure 1** might be used
in the investigation described
on page 98.

> Remember you are required to use geological
> maps as part of your investigation.

....

.... **(2 marks)**

> **Guided**

2 The students investigating coastal processes in Swanage Bay decided to use random
sampling to select their data collection sites. Explain **one** advantage and **one**
disadvantage of using random sampling.

Random sampling is the least biased of the sampling techniques because it is not

subjective so each site ....

....

.... **(4 marks)**

3 Explain the difference between **qualitative** and **quantitative** fieldwork.

....

....

....

.... **(4 marks)**

Working with data

1   Study **Figure 1**.

The shingle is higher on the south side as this is the direction of longshore drift so the pebbles transported by the waves are deposited here.

Wooden groyne built at right angles to the coastline to trap sediment.

The shingle is lower on the north side as the groyne traps the pebbles being transported, therefore less deposition occurs.

Figure 1 An annotated photograph of a groyne, Brighton

> **Guided**

(a) Suggest **one** advantage of using this method to present data.

Photographs give a very clear visual impression of the general area and sites used

to collect data, and ....

.... **(2 marks)**

(b) Select **one** other method that could be used to display coastal processes fieldwork data. Explain **one** disadvantage of using this method.

Chosen method: ....

....

.... **(2 marks)**

2   Study **Figure 2**.

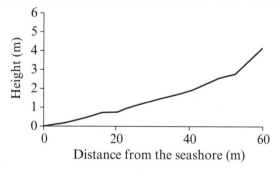

Figure 2 A beach profile

(a) Using **Figure 2**, describe how the beach height changes with distance from the seashore.

> Remember to include data in your answer.

....

.... **(2 marks)**

(b) Suggest **one** reason for the shape of the beach profile.

....

.... **(2 marks)**

Had a go ☐   Nearly there ☐   Nailed it! ☐

Formulating enquiry questions

This section is about Rivers fieldwork. For Coastal fieldwork, see page 98.

Guided 1 Identify the stages in the enquiry process by completing boxes 2, 4 and 6 in the diagram below. **(3 marks)**

| 1 Develop a question | → | 2 .... | → | 3 Process and present your data | → | 4 .... | → | 5 Make conclusions | → | 6 .... |

2 Study **Figure 1**.

Ordnance Survey Maps, © Crown copyright 2017, OS OS 100030901 and supplied by courtesy of Maps International.

Figure 1 Ordnance Survey map 1:25 000 of Dulverton, Somerset

A group of students intend to carry out fieldwork at a number of sites along the River Barle (**Figure 1**).

(a) Suggest an enquiry question based on river channel processes that they might investigate.

.... **(1 mark)**

(b) Suggest **one** reason why the River Barle is a suitable location to investigate your chosen enquiry question.

> Look carefully at the map.

....

.... **(2 marks)**

3 (a) State **one** geography theory or concept that might be investigated as part of this fieldwork.

.... **(1 mark)**

(b) Explain why the theory or concept named in (a) is relevant to this investigation.

....

....

....

.... **(4 marks)**

4 Study **Figure 1**. Name **one** man-made feature that might affect this fieldwork investigation.

.... **(1 mark)**

Methods and secondary data

1 (a) Identify which **one** of the following gives a definition of **secondary data**.

   ☐ **A** Data collected using sampling

   ☐ **B** Data collected very quickly

   ☐ **C** Data collected by someone else

   ☐ **D** Data collected directly during fieldwork
                                                                    (1 mark)

   (b) Study **Figure 1**.

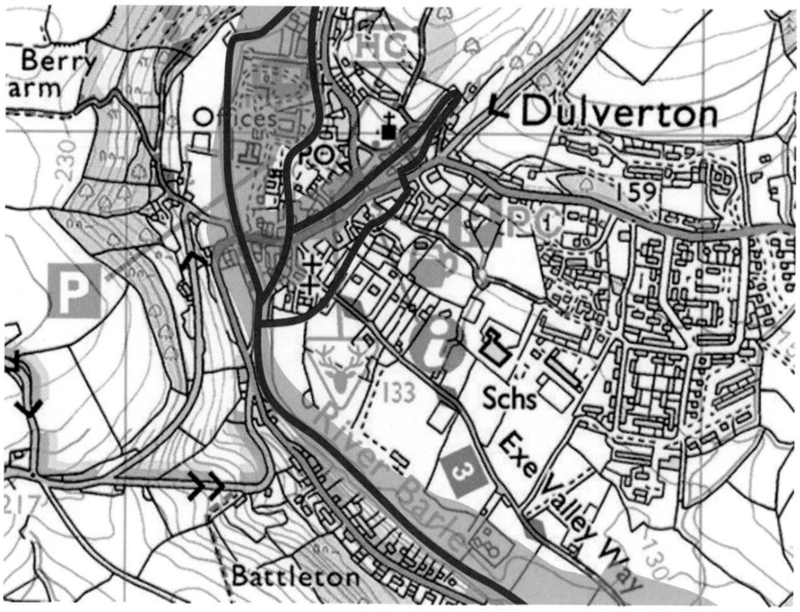

Ordnance Survey Maps, © Crown copyright 2017, OS 100030901 and supplied by courtesy of Maps International.

Figure 1 Flood risk map for Dulverton 1:25 000

Explain how the information shown on **Figure 1** might be used in the investigation described on page 101.

> Remember you are required to use flood risk maps as part of your investigation.

....

....   **(2 marks)**

Guided

2 The students investigating river channel processes in the Dulverton area decided to use random sampling to select their data collection sites. Explain **one** advantage and **one** disadvantage of using random sampling.

Random sampling is the least biased of the sampling techniques because it is not

subjective so each site ....

....   **(4 marks)**

3 Explain the difference between **qualitative** and **quantitative** fieldwork.

....

....

....

....   **(4 marks)**

Working with data

1 Study **Figure 1**.

A wall has been built as part of the flood defences along the river.

There is evidence that bed load has been deposited, during a period of high flow.

River is not full, indicating a period of low flow conditions.

Figure 1 An annotated photograph of Cray Beck, Yorkshire

>**Guided**

(a) Suggest **one** advantage of using this method to present data.

Photographs give a very clear visual impression of the general area and sites used

to collect data, and .... **(2 marks)**

(b) Select **one** other method that could be used to display river channel fieldwork data. Explain **one** disadvantage of using this method.

Chosen method: ....

....

.... **(2 marks)**

2 Study **Figure 2**.

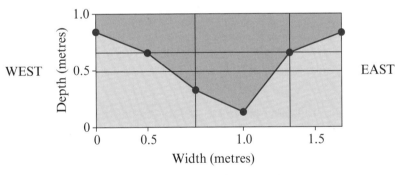

Figure 2 A river channel cross section

(a) Using **Figure 2**, describe how the river channel cross section changes with distance from the west bank.

> Remember to include data in your answer.

....

.... **(2 marks)**

(b) Suggest **one** reason for the shape of the river channel.

....

.... **(2 marks)**

Formulating enquiry questions

This section is about Urban fieldwork. For Rural fieldwork, see page 107.

1 (a) State the stage of the enquiry process that includes case studies and theories that help with explanations.

.... **(1 mark)**

Guided

(b) Explain what should be included in the evaluation stage of the enquiry process.

The evaluation stage should include a review of the data collection methods

....

.... **(2 marks)**

2 Study **Figure 1**.

Figure 1 A street in an inner city area

Some students intended to carry out fieldwork in the street shown in **Figure 1** to investigate the enquiry question: 'How does the quality of the urban environment vary along a transect in an inner city area?'

(a) Suggest a method they might use to collect quantitative data.

.... **(1 mark)**

(b) Explain the factors you would consider when designing a recording sheet to collect this data.

....

....

.... **(3 marks)**

(c) Explain **one** reason why the street shown in **Figure 1** might not be a suitable location to investigate this enquiry question.

....

.... **(2 marks)**

Methods and secondary data

Some students intended to investigate the quality of the urban environment in Fenham, an inner city area of Newcastle.

1 (a) State a qualitative fieldwork method they might use to collect this information.

.... **(1 mark)**

> **Guided**

(b) Explain **one** advantage and **one** disadvantage of using stratified sampling when collecting this information.

One advantage of using stratified sampling when carrying out a questionnaire

survey is that each age group within the population is properly represented, so

this gives an accurate ....

.... **(4 marks)**

2 Study **Figure 1**.

			% of total population of the ward
People and households	Total population	10 954	—
	Total households	4441	—
	Average household size	2.4	—
	Population aged 0–4 (early years)	740	6.8
	Population aged 5–14 (school years)	1271	11.6
	Population aged 15–24 (transition)	1952	17.8
	Population aged 25–64 (working age)	5390	49.2
	Population aged 65 and over (later life)	1601	14.6

Figure 1 2011 census data for Fenham ward, Newcastle

(a) Identify the type of data shown in **Figure 1**.

☐ **A** Primary and secondary   ☐ **C** Secondary and tertiary

☐ **B** Primary   ☐ **D** Secondary   **(1 mark)**

(b) Explain how the students could use the data in **Figure 1** to collect representative information.

....

.... **(2 marks)**

3 The students carried out their data collection on a Monday in September, between 10:00 am and 12:30 pm. Explain why the information they collected would not be representative.

....

....

....

.... **(4 marks)**

Had a go ☐ Nearly there ☐ Nailed it! ☐

Working with data

Some students intended to investigate the quality of the urban environment in Fenham, an inner city area of Newcastle.

1 Study **Figure 1**.

Question: Is litter a problem in Fenham?		
Age group	Number of replies Yes	Number of replies No
0–4	0	0
5–14	2	1
15–24	6	0
25–64	7	18
Over 65	3	0

Figure 1 A table showing the replies to a question about the quality of the urban environment

A student planned to construct a scattergraph to show the data shown in **Figure 1**.

(a) Suggest why a scattergraph is an inappropriate method to show this data.

You should be able to explain the advantages and disadvantages of a number of data presentation methods.

....

.... **(2 marks)**

(b) Suggest a graphing method that might be used to show this data.

.... **(1 mark)**

(c) Explain **one** disadvantage of using the method you have named in (b).

....

.... **(2 marks)**

Guided

2 Suggest **two** additional questions that the students might ask as part of their investigation of the environmental quality of Fenham ward.

(i) Are there enough green spaces in Fenham ward?

(ii) .... **(2 marks)**

3 You have carried out fieldwork in an urban area. Evaluate the methods you used to collect your data.

....

....

....

....

.... **(8 marks)**

Continue your answer on your own paper. You should aim to write approximately one side of A4.

Formulating enquiry questions

This section is about Rural fieldwork. For Urban fieldwork, see page 104.

1 (a) State the stage of the enquiry process that includes case studies and theories that help with explanations.

.... **(1 mark)**

Guided (b) Explain what should be included in the evaluation stage of the enquiry process.

The evaluation stage should include a review of the data collection methods

....

.... **(2 marks)**

2 Study **Figure 1**.

Figure 1 Llanbedr, a rural village in North Wales

Some students intended to carry out fieldwork in the village shown in **Figure 1** to investigate the enquiry question: 'Do traffic flows vary during the day in a rural settlement?'

(a) Suggest a method they might use to collect quantitative data.

.... **(1 mark)**

(b) Explain the factors you would consider when designing a recording sheet to collect this data.

....

....

.... **(3 marks)**

(c) Explain **one** reason why the village shown in **Figure 1** might not be a suitable location to investigate this enquiry question.

....

.... **(2 marks)**

Had a go ☐ Nearly there ☐ Nailed it! ☐

Methods and secondary data

Some students intended to investigate the quality of the rural environment in Llanbedr, the village shown in Figure 1 on page 107.

1 (a) State a qualitative fieldwork method they might use to collect this information.

.... **(1 mark)**

Guided

(b) Explain **one** advantage and **one** disadvantage of using stratified sampling when collecting this information.

One advantage of using stratified sampling is that each age group within the population is properly represented, so this gives an accurate ....

....

.... **(4 marks)**

2 Study **Figure 1**.

Age group	Number in each age group	% of total population
0–4	20	7.7
5–14	24	9.2
15–24	23	8.8
25–64	130	50
Over 65	63	24.3

Figure 1 2011 census data for Llanbedr

(a) Identify the type of data shown in **Figure 1**.

☐ **A** Primary and secondary

☐ **B** Primary

☐ **C** Secondary and tertiary

☐ **D** Secondary **(1 mark)**

(b) Explain how the students could use the data in **Figure 1** to collect representative information.

....

.... **(2 marks)**

3 The students carried out their data collection on a Monday in September, between 10:00 am and 12:30 pm. Explain why the information they collected would not be representative.

....

....

....

.... **(4 marks)**

Working with data

Some students intended to investigate the quality of the rural environment in Llanbedr, the village shown in Figure 1 on page 107.

1 Study **Figure 1**.

Question: Is litter a problem in Llanbedr?		
Age group	Number of replies Yes	Number of replies No
0–4	0	0
5–14	2	1
15–24	6	0
25–64	7	18
Over 65	3	0

Figure 1 A table showing the replies to a question about the quality of the rural environment

A student planned to construct a scattergraph to show the data shown in **Figure 1**.

(a) Suggest why a scattergraph is an inappropriate method to show this data.

> You should be able to explain the advantages and disadvantages of a number of data presentation methods.

....

.... **(2 marks)**

(b) Suggest a graphing method that might be used to show this data.

.... **(1 mark)**

(c) Explain **one** disadvantage of using the method you have named in (b).

....

.... **(2 marks)**

> **Guided**

2 Suggest **two** additional questions that the students might ask as part of their investigation of the environmental quality of Llanbedr.

(i) Are there enough green spaces in Llanbedr?

(ii) .... **(2 marks)**

3 You have carried out fieldwork in a rural area. Evaluate the methods you used to collect your data.

....

....

....

....

.... **(8 marks)**

Continue your answer on your own paper. You should aim to write approximately one side of A4.

Had a go ☐ Nearly there ☐ Nailed it! ☐

Consumption and environmental challenges

1 (a) Explain the meaning of the term **overpopulated**.

....

.... **(2 marks)**

Guided

(b) Study **Figure 1**.

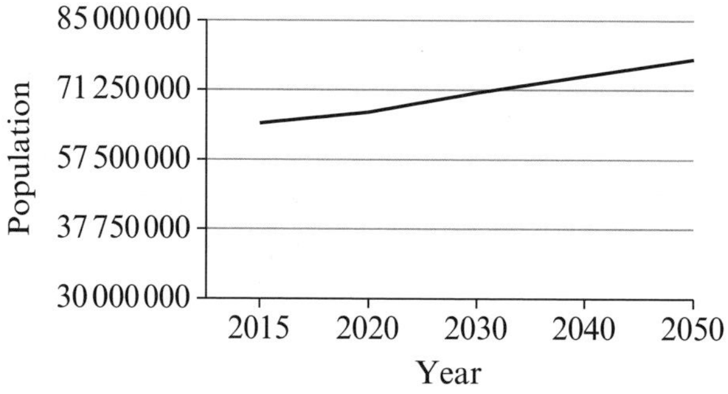

Figure 1 UK projected population growth 2015–2050

Explain the implications on resource consumption of the population growth shown in **Figure 1**. Include data from **Figure 1** in your answer.

The UK's population is predicted to increase steadily from 65 million in 2015 to more than 77 million in 2050. This is an increase of ....

....

....

....

.... **(4 marks)**

(c) Identify **two** pressures caused by population growth on ecosystems.

☐ **A** Reduction in groundwater supplies

☐ **B** Increased biodiversity

☐ **C** Loss of habitats

☐ **D** Reduction in atmospheric pollution **(2 marks)**

2 (a) Name a sustainable transport system.

.... **(1 mark)**

(b) Explain **two** reasons why the UK is developing a range of sustainable transport systems.

Sustainable transport allows people to travel but does not adversely affect present or future humans or ecosystems.

....

....

....

.... **(4 marks)**

Population and economic challenges

1 (a) State what is meant by the term **two-speed economy**.

.... **(1 mark)**

(b) Study **Figure 1**.

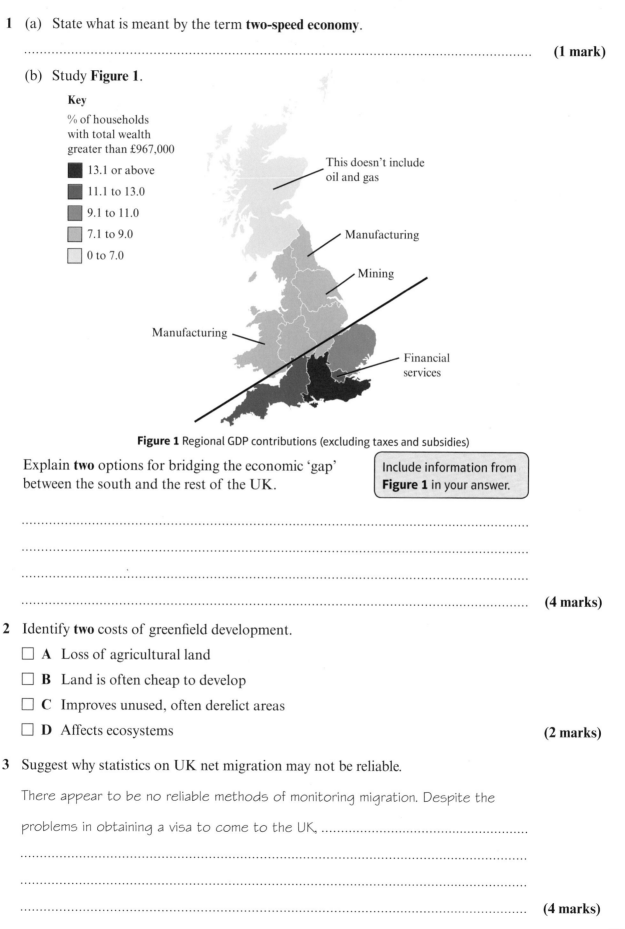

Key

% of households with total wealth greater than £967,000

■ 13.1 or above

■ 11.1 to 13.0

■ 9.1 to 11.0

■ 7.1 to 9.0

☐ 0 to 7.0

This doesn't include oil and gas

Manufacturing

Mining

Manufacturing

Financial services

Figure 1 Regional GDP contributions (excluding taxes and subsidies)

Explain **two** options for bridging the economic 'gap' between the south and the rest of the UK.

> Include information from **Figure 1** in your answer.

....

....

....

.... **(4 marks)**

2 Identify **two** costs of greenfield development.

☐ **A** Loss of agricultural land

☐ **B** Land is often cheap to develop

☐ **C** Improves unused, often derelict areas

☐ **D** Affects ecosystems **(2 marks)**

⟩ **Guided** ⟩ 3 Suggest why statistics on UK net migration may not be reliable.

There appear to be no reliable methods of monitoring migration. Despite the

problems in obtaining a visa to come to the UK, ...

....

....

.... **(4 marks)**

Landscape challenges

1 (a) Identify **two** UK National Parks.

☐ **A** Dartmoor

☐ **B** Epping Forest

☐ **C** Cairngorms

☐ **D** North Downs   **(2 marks)**

(b) Explain **two** approaches to the conservation of UK National Parks.

....

....

....

.... **(4 marks)**

2 Study **Figure 1**.

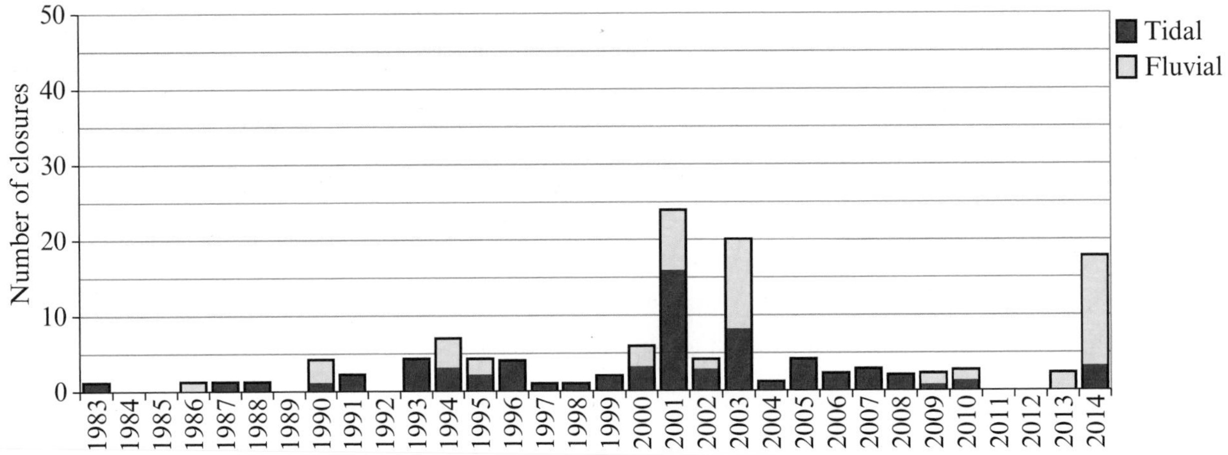

Figure 1 Frequency of closures of the Thames Barrier 1983-2014

⟩ **Guided** ⟩

(a) Describe the trend of Thames Barrier closures between 1983 and 2014.
Use data from **Figure 1** in your answer.

> **Fluvial flooding** means the same as river flooding.

There has been a marked increase in the number of closures. Between 1983 and

1989 there was only one closure a year or less, but after 1990 this increased to

....

.... **(3 marks)**

(b) Explain **two** different approaches to managing the coastal flood risk.

....

....

....

.... **(4 marks)**

Climate change challenges

1   Study **Figure 1**.

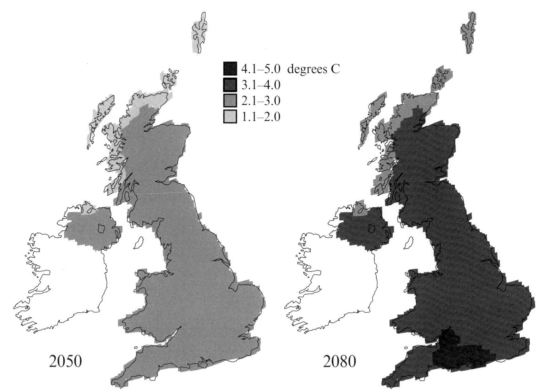

■ 4.1–5.0 degrees C
■ 3.1–4.0
■ 2.1–3.0
☐ 1.1–2.0

2050   2080

Figure 1 Projected changes in the UK's mean summer temperature for 2050 and 2080

(a)   Identify the predicted highest mean summer temperature increase by 2080.

☐ **A** + 2 °C in northern Scotland

☐ **B** + 4 °C in southern England

☐ **C** + 3 °C in central England

☐ **D** + 5 °C in southern England   **(1 mark)**

(b)   Suggest how global climate change might affect precipitation in the UK.

....

....   **(2 marks)**

> **Guided**   2   (a)   State **two** local-scale responses to climate change.

(i) People can use public transport to reduce their carbon footprint.

(ii) ....

(b)   Explain the responses to climate change at a national scale in the UK.

....

....

....

....   **(4 marks)**

Paper 3 (i)

> You should answer either Question 1: Rivers or Question 2: Coasts.

⟩ Guided ⟩   1   Study **Figure 1**.

River Clyne, Swansea, channel characteristics investigation

Information River length: 6.5 km   Course: SE through Clyne Country Park   Source: springs   Mouth: Blackpill

Conclusions
- River width at site 1 – 3.2 metres, increasing to 5.8 metres at site 3 due to tributaries joining the river
- River depth increases from 22 cm at site 1 to 68 cm at site 3 because of erosion
- River load gets smaller and rounder downstream due to erosion

Overall the River Clyne changes downstream as predicted by the Bradshaw model.

Figure 1 Student's fieldwork conclusions

Evaluate the student's conclusions.

The student only investigated three sites. Therefore ....

....

....

....

....

.... **(8 marks)**

> Continue your answer on your own paper. You should aim to write about $\frac{3}{4}$ of a side of A4.

⟩ Guided ⟩   2   Study **Figure 2**.

West Wittering Beach, Chichester, coastal process investigation

Information An extensive sand and shingle beach with coastal protection

Conclusions
- Beach profile 1 (mean angle 9⁰) steeper than beach profile 3 (6⁰) due to sediment trapped by the groyne
- Beach sediment increased in size from sand near the sea to shingle at the sea end of the profile because the shingle beach is a high-energy storm beach
- Shingle was mostly made up of sedimentary rocks from cliffs to the west of West Wittering

Overall the beach at West Wittering shows that longshore drift is the most important coastal process.

Figure 2 Student's fieldwork conclusions

Evaluate the student's conclusions.

The student only investigated three sites. Therefore ....

....

....

....

....

.... **(8 marks)**

> Continue your answer on your own paper. You should aim to write about $\frac{3}{4}$ of a side of A4.

Paper 3 (ii)

Guided

1   Study **Figure 1 on page 131**, which shows possible impacts of climate change on the UK.

Use information from **Figure 1** and knowledge and understanding from the rest of your geography course.

Discuss the view that climate change will have an entirely negative impact on the people and landscapes of the UK.

The content for Assessment Objective 2, which is worth 4 marks, might include the following explanations for climate change.

- Variations in the amount of solar energy reaching Earth.
- Changes in how much incoming solar radiation is reflected back into space.
- Variations in the greenhouse effect which affect the amount of heat retained by Earth's atmosphere.

For Assessment Objective 3, which is worth 4 marks, you would need to:

- weigh up the relative importance of positive and negative impacts of climate change on the people and landscapes of the UK
- make a judgement based on evidence that includes facts and figures.

For Assessment Objective 4, which is worth 4 marks, you need to include information from Figure 1, such as:

- Scotland will have increased flooding and loss of ecosytems
- new species will spread into sea areas, increasing fishing opportunities.

For SPGST, which is worth 4 marks, you need to:

- spell (especially geographical terms) and punctuate accurately
- make sure you follow the rules of grammar
- use a wide range of geographical terms.

....

....

....

....

....

....

....

....

....

....

....

(12 marks plus 4 marks for SPGST)

Continue your answer on your own paper. You should aim to write about a side of A4.

Atlas and map skills

In the exam, there will not be a separate section testing your geographical, mathematics and statistics skills. Instead, you might need to use these skills in any of the three exam papers, when answering questions about the topics you have studied. The questions on pages 116 to 130 are designed to help you practise these skills. Not all of the questions are in the style of the questions you will find in the exam.

1   Study **Figure 1**.

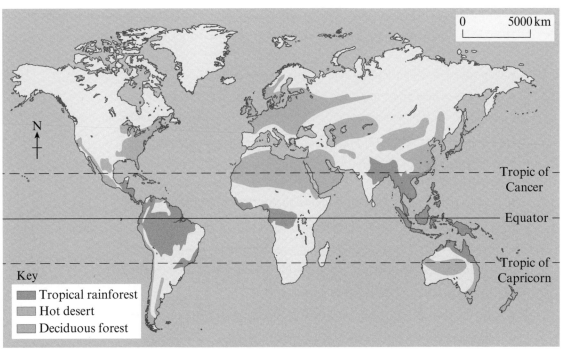

Figure 1 The distribution of three global ecosystems

(a)   Identify **two** terms used to describe patterns on maps.

☐ **A** Dispersed

☐ **B** Hilly

☐ **C** Wooded

☐ **D** Linear

> You are expected to use geographical terms.

(2 marks)

Guided

(b)   Describe the distribution of deciduous forest ecosystems.

Deciduous forests and woodlands are mostly located north of the Tropic of Cancer.

They form a dispersed linear band ....

....

....   **(3 marks)**

2   (a)   **Figure 1** shows the distribution of three ecosystems. State **two** other types of distribution shown on atlas maps.

(i) ....   (ii) ....   **(2 marks)**

(b)   Physical maps show relief. Explain what is meant by the term **relief**.

....

....   **(2 marks)**

Types of map and scale

1 (a) Identify what is meant by **the scale of a map**.

   ☐ **A** The size of an area on a map compared to real life

   ☐ **B** A map drawn to half size

   ☐ **C** A map drawn the same size as real life

   ☐ **D** A map drawn to twice real size   **(1 mark)**

> **Guided**

   (b) State the **two** scales most frequently used for Ordnance Survey maps.

   (i) ....   (ii) 1:50 000   **(2 marks)**

2 Study **Figure 1**.

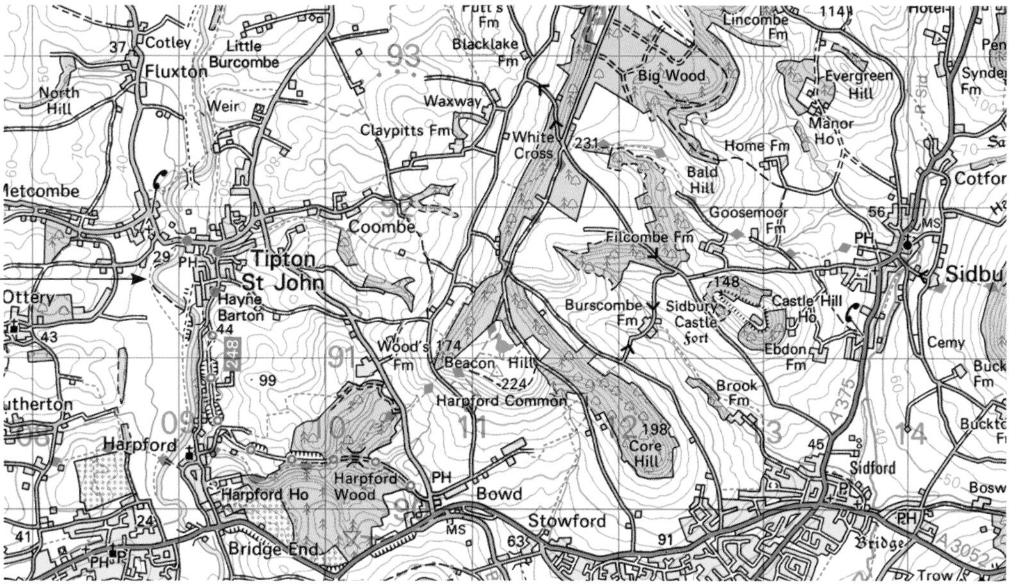

Figure 1 Ordnance Survey map scale 1:50 000

Using **Figure 1**, calculate the distance along the River Otter to the nearest $\frac{1}{2}$ km.

   ☐ **A** 4 km

   ☐ **B** 16 km

   ☐ **C** 6 km

   ☐ **D** 10 km   **(1 mark)**

> Use a piece of string to help you work out winding distances, such as along rivers.

3 (a) Explain what is meant by the term **isoline.**

   ....

   ....   **(2 marks)**

   (b) Name **one** type of map that uses isolines to show information.

   ....   **(1 mark)**

Using and interpreting images

1 Study **Figure 1**.

Figure 1 Photograph of the Hayle estuary, Cornwall

(a) Identify the type of photograph shown in **Figure 1**.

☐ **A** Ground level   ☐ **C** Vertical aerial

☐ **B** Oblique aerial   ☐ **D** Satellite   **(1 mark)**

(b) Suggest **one** advantage of using this type of photograph.

.... **(1 mark)**

Guided 2 (a) State **one** disadvantage of using ground-level aerial photographs.

The detail in the background may not .... **(1 mark)**

(b) Explain how satellite images can be used.

....

.... **(2 marks)**

3 Study **Figure 2**.

Figure 2 Photograph of a coastal landscape

Suggest **one** type of erosion taking place in the coastal area shown in **Figure 2**.

> You should be able to explain what is happening in photographs.

.... **(1 mark)**

Had a go ☐   Nearly there ☐   Nailed it! ☐

Sketch maps and annotations

1 Explain **one** use of a simple sketch map.

....

.... **(2 marks)**

Guided **2** Study **Figure 1**.

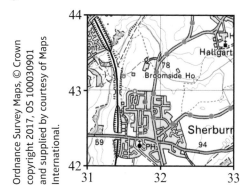   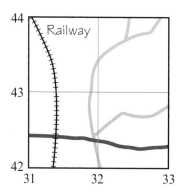

Figure 1 Ordnance Survey 1:50 000 map of the Sherburn area

Draw and label a sketch map to the right of **Figure 1** to show the following features:

> Use the grid lines to help you to be accurate.

~~the railway~~   ~~roads~~   the church   Broomside House   Sherburn   **(5 marks)**

3 Study **Figure 2**.

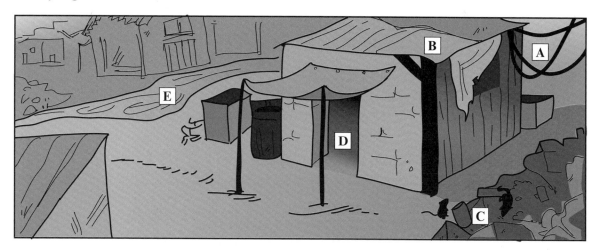

Figure 2 A sketch of a squatter settlement

The **labels A–E** on **Figure 2** match the descriptions in the grid below.
Put the letters A–E next to the correct description in the grid.

1 Lack of access to a proper water supply	
2 Rubbish is often left at the roadside which attracts rats	
3 Some shanty towns have access to services such as electricity	
4 Very small roomed accommodation which leads to overcrowding	
5 Houses are often made out of lots of different materials	

(5 marks)

Had a go ☐ Nearly there ☐ Nailed it! ☐

Physical and human patterns

1 Explain **one** method that can be used to describe patterns.

....

.... **(2 marks)**

2 Study **Figure 1**.

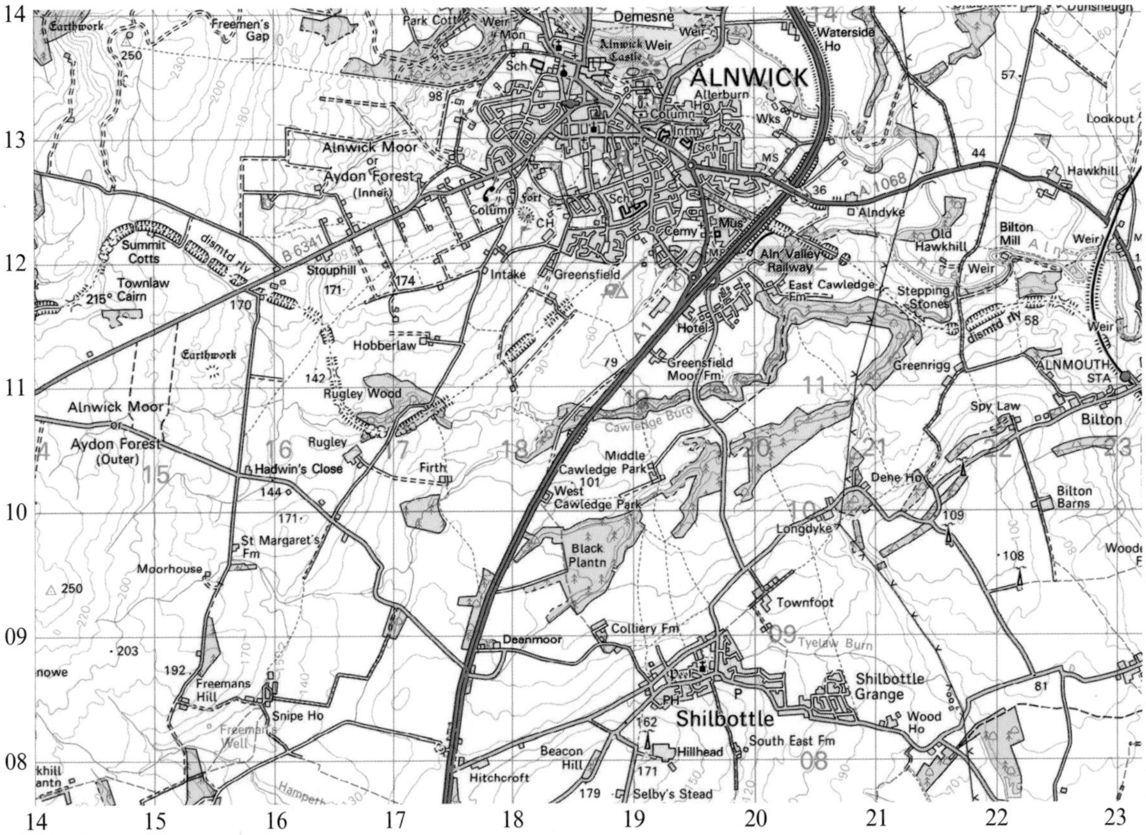

Ordnance Survey Maps, © Crown copyright 2017, OS 100030901 and supplied by courtesy of Maps International.

Figure 1 Ordnance survey map of the Alnwick area, 1:50 000

(a) Describe the shape of Alnwick, a settlement shown on the OS map extract in **Figure 1**. Use map evidence in your answer.

> Include road numbers and compass points in your answer.

....

....

.... **(3 marks)**

Guided

(b) Explain why it would be difficult to expand Alnwick. Use map evidence in your answer.

It would be difficult to expand Alnwick because of the ....

road to the .... of the settlement ....

....

.... **(4 marks)**

Land use and settlement shapes

1   Explain how Ordnance Survey maps can be used to explain land use.

....

.... **(2 marks)**

2   (a)   Identify the description of a dispersed settlement.

☐ **A**   Clustered or grouped together

☐ **B**   In a line along a road

☐ **C**   Spread out

☐ **D**   Contains shops and offices   **(1 mark)**

(b)   Study **Figures 1a** and **1b**.

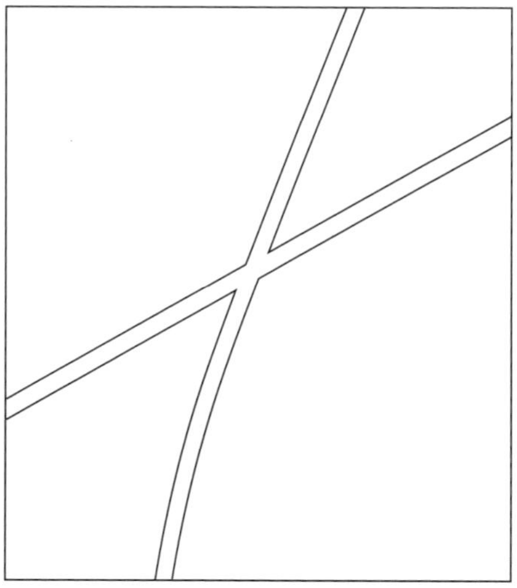

Figure 1a 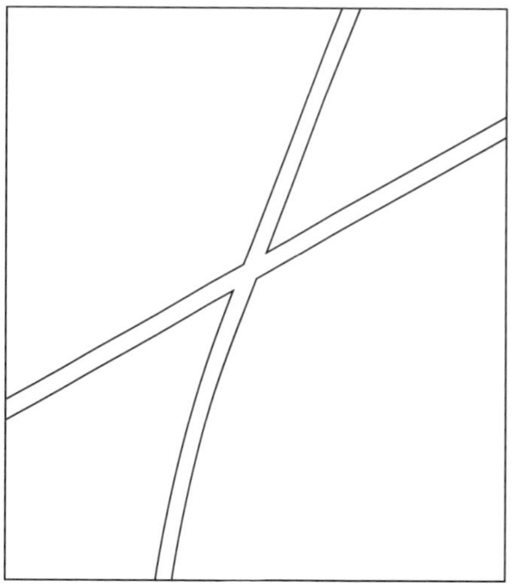   **Figure 1b**

(i)   In **Figure 1a**, draw the shape of a nucleated settlement.   **(1 mark)**

(ii)   In **Figure 1b**, draw the shape of a linear settlement.   **(1 mark)**

3   Look back at the map extract on page 117.

(a)   Describe the shape of Sidford in grid square 1390.

....

.... **(2 marks)**

> **Guided**

(b)   Describe the physical and human land uses from grid square 1290 to grid square 1390. Use map evidence in your answer.

> The best answers will include specific references to the map.

The land use is very rural. Brook Farm is located to the ....

....

....

.... **(4 marks)**

Human activity and OS maps

Look at the map extract of Alnwick on page 120.

1 (a) State **one** piece of map evidence in grid square 1912 that indicates Alnwick attracts tourists.

.... **(1 mark)**

(b) Suggest **one** other piece of map evidence that indicates Alnwick is a popular tourist destination.

.... **(1 mark)**

2 (a) Identify **one** land use found in grid square 2010.

☐ **A** Houses

☐ **B** Deciduous woodland

☐ **C** Coniferous woodland

☐ **D** Factories　　　　　　　　　　　　　　　　　　　　　　　　**(1 mark)**

(b) State **one** land use in grid square 1813 that indicates Alnwick is an urban area.

.... **(1 mark)**

> **Guided**

3 'Shilbottle is a rural settlement.' Explain this statement using map evidence to support your answer.

Shilbottle is surrounded by a rural landscape. There are many farms around the area,

which show it is rural agricultural area. For example, South East Farm located to

the south east of Shilbottle. ....

....

.... **(4 marks)**

4 In the box below, draw and label a sketch to show how an industrial area would be represented on an OS map.　　　　　　　　　　　　　　　　　　　　　　**(3 marks)**

Map symbols and direction

> **Guided**

1 Draw the OS map symbols (1:50 000) for the following features.

Golf course   🚩   Church with a tower ☐   Spot height ☐

Nature Reserve ☐   Clubhouse ☐   Coniferous wood ☐

(2 marks)

2 Study **Figure 1**.

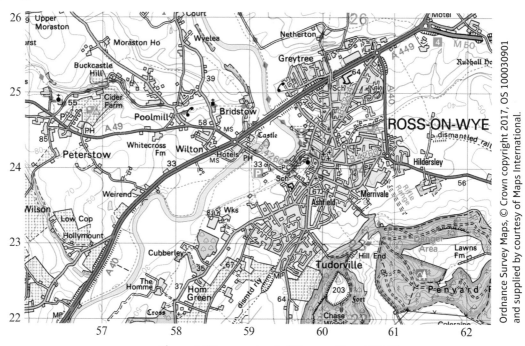

Figure 1 OS map extract of Ross-on-Wye, 1:50 000

Ordnance Survey Maps, © Crown copyright 2017, OS 100030901 and supplied by courtesy of Maps International.

(a) State the general direction followed by the A449.

.... **(1 mark)**

(b) State the land use found in the south-east corner of the map.

> Remember to use the key of a 1:50 000 map extract to check map symbols.

.... **(1 mark)**

(c) State the road number of the major road found in the north-east corner of the map.

.... **(1 mark)**

Grid references and distances

Look at the OS map extract on page 117.

1 (a) State the type of woodland found in grid square 1292.

.... **(1 mark)**

(b) State the six-figure grid reference for the nature reserve in the centre of the extract.

.... **(1 mark)**

(c) To the nearest $\frac{1}{2}$ km, work out the straight line distance from the public house in grid square 1391 to the public house in grid square 0991.

> Read the question carefully.

.... **(1 mark)**

(d) Identify the farm in grid square 1092.

☐ **A** Home Farm

☐ **B** Claypitts Farm

☐ **C** Brook Farm

☐ **D** Goosemoor Farm   **(1 mark)**

2 There are two bridges located on the River Otter (from grid square 0990 to 0992). Identify the six-figure grid reference of the most northerly bridge.

☐ **A** 099921

☐ **B** 095926

☐ **C** 091923

☐ **D** 095925   **(1 mark)**

3 (a) State the total winding distance of the River Otter to the nearest $\frac{1}{2}$ km.

.... **(1 mark)**

(b) Identify the map symbol found at grid square 137912.

.... **(1 mark)**

(c) State the six-figure grid reference for the highest point in grid square 1190.

.... **(1 mark)**

> Guided

(d) State the six-figure grid reference of the telephone which appears to be in grid square 0892.

087.. **(1 mark)**

Cross-sections and relief

Guided 1 Draw lines to match up the following contour patterns (aerial view) to the cross-sectional shape of the land.

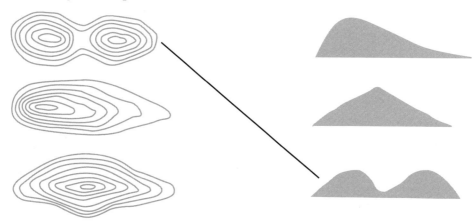

(3 marks)

2 Look at the OS map on page 117.

(a) Identify the maximum height of the land to the east of the River Otter.

☐ **A** 109 m

☐ **B** 95 m

☐ **C** 87 m

☐ **D** 99 m

(1 mark)

(b) State the maximum height of the land to the west of the River Otter.

.... **(1 mark)**

3 (a) Draw a cross-sectional diagram of the contour pattern below.

> Use a sharp pencil for all graphical skills.

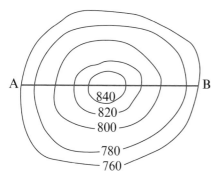

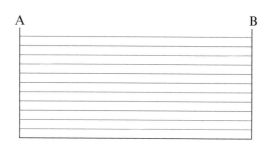

(4 marks)

(b) Describe the relief shown by your completed cross-section contour pattern.

....

....

.... **(3 marks)**

Graphical skills 1

Guided **1** The graph below shows the data collected during a three-minute traffic survey. Plot the data in the table to complete the graph.

> Be accurate – use the guide lines on the graph.

(a)

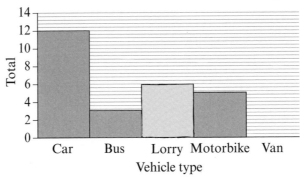

| Lorry | 6 |
| Van | 4 |

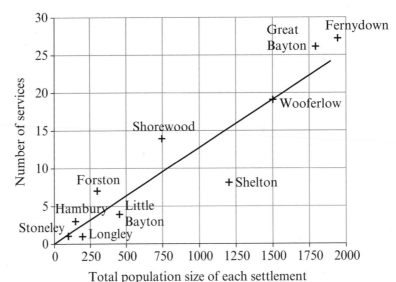

Figure 1 Data from three-minute traffic survey

(2 marks)

(b) A student decided to present the traffic data as a line graph. Explain why this graphical technique is inappropriate for the data collected.

....

....

.... **(3 marks)**

2 Study **Figure 2**.

Figure 2 Sample scattergraph

Describe the correlation shown in the scattergraph. Use evidence from **Figure 2** in your answer.

....

....

.... **(3 marks)**

Graphical skills 2

1 Study **Figure 1**.

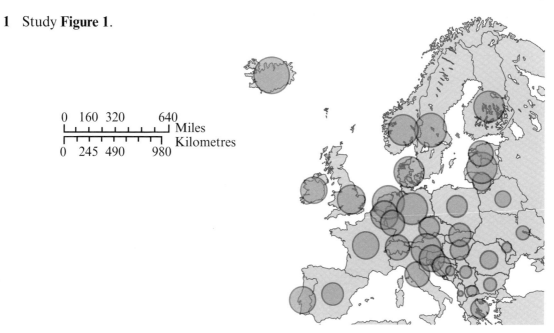

Figure 1 Internet users in the European Union

Suggest **one** disadvantage of using this method to present information.

....

.... **(2 marks)**

2 Study **Figure 2**.

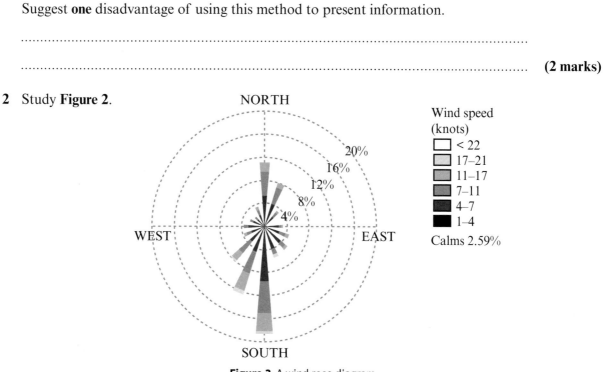

Figure 2. A wind rose diagram

⟩**Guided**⟩   (a) State what this graphical method shows.

A wind rose diagram shows wind direction .... **(2 marks)**

(b) Explain what this data shows about the wind direction for this particular area.

> You need to make sure you can recognise patterns on a range of graphs.

....

....

.... **(3 marks)**

Graphical skills 3

Guided   **1**   (a)   State **two** characteristics of a population pyramid.

(i) Data is shown for males and females

(ii) .... **(2 marks)**

(b)   Study **Figure 1**.

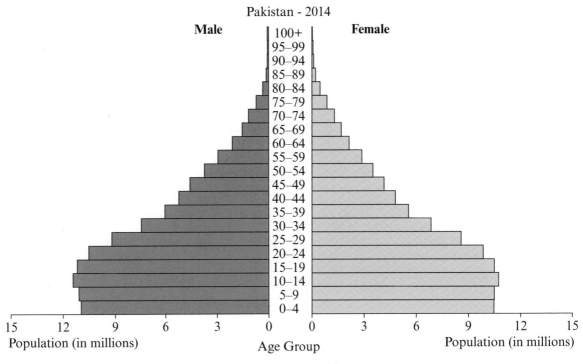

Figure 1 Population pyramid for Pakistan

Suggest the probable level of development in Pakistan.

.... **(1 mark)**

(c)   Explain your answer to question (b).

> Use evidence from the population pyramid to support your answer.

....

....

2   Suggest **one** advantage and **one** disadvantage of using choropleth maps.

....

.... **(2 marks)**

3   Describe why flow diagrams are an effective method to present data.

....

....

.... **(3 marks)**

Numerical and statistical skills 1

⟩ **Guided** ⟩

1 The total area of the UK is 243 610 km². The area of land used for agriculture is 170 527 km².
Calculate the percentage of land in the UK used for agriculture.

Show your working.

170 527 km² ÷ 243 610 km² = ... **(2 marks)**

2 Study **Figure 1**.

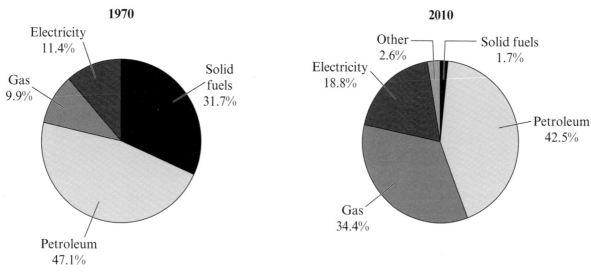

Figure 1 UK energy consumption 1970 and 2010

(a) Calculate the percentage increase in gas consumption between 1970 and 2010.

.... **(1 mark)**

(b) Calculate the percentage decrease in solid fuel consumption between 1970 and 2010.

.... **(1 mark)**

3 The dependency ratio for a country is calculated by the following formula:

$$\frac{\text{Non-economically active} \times 100}{\text{Economically active}}$$

Calculate the dependency ratio for the UK using the following information (2011 Census figures).

Non-economically active = 21 478 million   Economically active = 41 704 million

= .... **(2 marks)**

Numerical and statistical skills 2

> **Guided**

1 Study **Figure 1**.

Pebble long axis (cm)	First calculation	Second and third calculation
24		
22		
19		Upper quartile (2 above, 2 below)
18		
16		
12	Median	
10		
9		
7		
5		
3		

Figure 1 Table showing information about pebbles measured as part of a beach process investigation

(a) State what is meant by the term **median value**.

.... **(1 mark)**

(b) Calculate the mean of the pebble long axes.

.... **(1 mark)**

(c) State the formula for calculating the lower quartile range.

.... **(1 mark)**

(d) Label **lower quartile** in the relevant place on the table.   **(1 mark)**

2 (a) Calculate the interquartile range for the data in **Figure 1**.

[Show your working.]

....

.... **(2 marks)**

(b) Suggest what information is given by the calculated interquartile range (IQR).

....

.... **(2 marks)**

3 Identify the meaning of the term **modal class**.

☐ **A** The number in the middle of a group

☐ **B** The group that has the highest frequency

☐ **C** The average of the numbers in a group

☐ **D** The size of a group   **(1 mark)**

Resource for Paper 3 (ii)

You will need to refer to **Figure 1** below when you answer the 12-mark question on page 115.

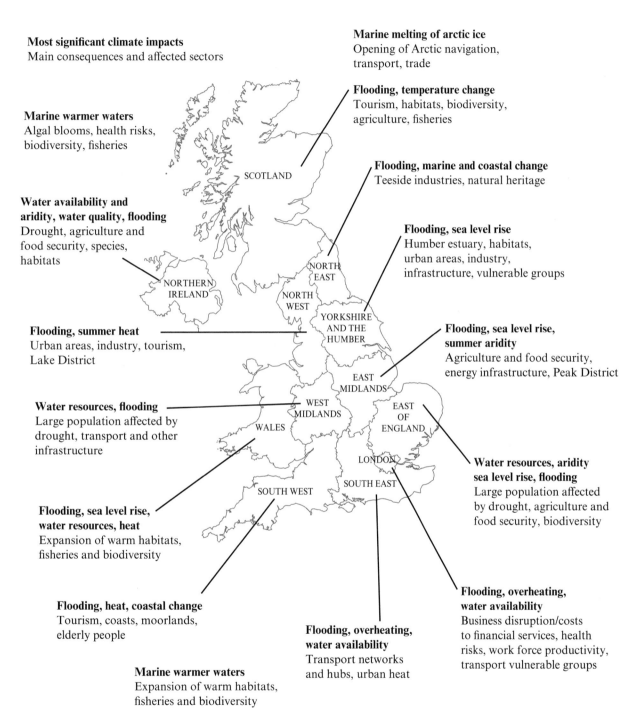

Most significant climate impacts
Main consequences and affected sectors

Marine melting of arctic ice
Opening of Arctic navigation, transport, trade

Flooding, temperature change
Tourism, habitats, biodiversity, agriculture, fisheries

Marine warmer waters
Algal blooms, health risks, biodiversity, fisheries

Flooding, marine and coastal change
Teeside industries, natural heritage

Water availability and aridity, water quality, flooding
Drought, agriculture and food security, species, habitats

Flooding, sea level rise
Humber estuary, habitats, urban areas, industry, infrastructure, vulnerable groups

SCOTLAND

NORTH EAST

NORTHERN IRELAND

NORTH WEST

YORKSHIRE AND THE HUMBER

Flooding, summer heat
Urban areas, industry, tourism, Lake District

Flooding, sea level rise, summer aridity
Agriculture and food security, energy infrastructure, Peak District

EAST MIDLANDS

Water resources, flooding
Large population affected by drought, transport and other infrastructure

WEST MIDLANDS

EAST OF ENGLAND

WALES

LONDON

SOUTH EAST

Water resources, aridity sea level rise, flooding
Large population affected by drought, agriculture and food security, biodiversity

SOUTH WEST

Flooding, sea level rise, water resources, heat
Expansion of warm habitats, fisheries and biodiversity

Flooding, overheating, water availability
Business disruption/costs to financial services, health risks, work force productivity, transport vulnerable groups

Flooding, heat, coastal change
Tourism, coasts, moorlands, elderly people

Flooding, overheating, water availability
Transport networks and hubs, urban heat

Marine warmer waters
Expansion of warm habitats, fisheries and biodiversity

Figure 1 Possible impacts of climate change on the UK

ANSWERS

Where an example answer is given, this is not necessarily the only correct response. In most cases there is a range of responses that can gain full marks.

COMPONENT 1: THE PHYSICAL ENVIRONMENT
Changing UK landscapes

1. Main UK rock types

1 (a)

Letter	Rock name	Rock type
Y	Granite	Igneous
Z	Chalk	Sedimentary

(b) Granite is formed of crystals
(c) **C** Formed by heat and pressure
(d) Sandstone

2 Granite is impermeable. This means <u>granite landscapes tend to drain badly because rain and drainage water cannot percolate as granite is crystalline.</u>

2. Upland and lowland landscapes

1 The Tees/Exe line
2 (a) **B** Metamorphic and igneous
(b) Sedimentary rocks, such as chalk, are relatively soft and are not made up of crystals. Therefore these are more easily broken down by weathering and erosion than are igneous and metamorphic rocks. This results in sedimentary rocks forming the low hills and lowland basins south of the Tees/Exe line.
3 Diverging plate boundaries about 60 million years ago meant that the Atlantic Ocean began to form. Rising lava <u>cooled and contracted to form polygonal basalt shapes.</u>

3. Physical processes

1 (a) **D** Weathering (b) The corrie containing Stickle Tarn was formed by glacial <u>erosion. A glacier carved out a semi-circular depression as it moved downhill.</u>
2 The lowland landscape in the photograph was formed by erosional and depositional processes operating together. As the rivers meander across their floodplains, erosion occurs. This produces silt and other sediment, which is transported by the river. When the rivers lose energy – for example, when flooding takes place or on the inside of meander bends – this sediment is deposited to form low-lying river floodplains.

4. Human activity

1 (a) Straight drainage ditches constructed (b) Forestry can involve planting trees in straight lines. This changes the appearance of the landscape because natural woodland forms irregular patterns.
2 There are different types of farming in the UK's upland and lowland landscapes. Steep slopes and poor soils in upland areas mean that <u>these are used for rough grazing. Lowland landscapes may have straight drainage ditches and large, regular shaped fields.</u>
3 (a) **A** Risk of flooding (b) Building settlements such as Shrewsbury means that the natural landscape has been changed to a human landscape. Houses and other buildings are tightly packed together along roads in the centre of the meander loop.
4 The density of settlements is much lower in upland areas. Settlements in upland areas tend to be small, scattered and located in valleys. Upland areas have steeper slopes, which are difficult to build on, so settlements are restricted to the flatter valley floors. The relief of upland areas makes it more difficult for settlements to expand than in lowland areas, so they tend to be villages or small market towns. The population density is lower in upland areas because there are fewer jobs available so there is less urbanisation, resulting in less impact on the landscape.

Coastal landscapes
5. Physical processes 1

1 Weathering
2 <u>Water enters cracks in the rock.</u> The water freezes and the cracks are widened. <u>Repeated freezing and thawing causes the rock to break apart.</u>
3 **D** Mass movement
4 One way the coast is eroded is by abrasion, which takes place when breaking waves throw sand, pebbles and boulders against the coast during storms, causing rocks to be broken down and removed. Another type of coastal erosion is hydraulic action, where the weight and energy of waves, particularly during a storm, erode the coast. As part of this process, waves compress air in joints in rocks, forcing them apart.

6. Physical processes 2

1 (a) Longshore drift (b) Backwash
2 The largest material, such as boulders, is rolled <u>along the seabed by traction. Smaller pebbles are bounced along the seabed by saltation.</u>
3 (a) Constructive (b) The load transported by a wave is deposited when a wave loses energy: for example, when the shallow gradient of a beach causes friction. The crest of the wave breaks towards the land creating a strong swash, which carries the material up the beach. As the constructive wave loses more energy, this material is deposited and the backwash is too weak to take it back to sea.

7. Influence of geology

1 (a) **X**: Discordant coast
Y: Headland
Z: Concordant coast
(b) A coastline made up of hard rocks will have cliffs that are high, steep and rugged, and landforms such as wave-cut platforms. However, a coastline made up of soft rocks will be less rugged and steep and have bays.
2 Destructive waves are <u>high-energy waves that cause coastal erosion.</u> However, constructive waves are low-energy waves that <u>cause coastal deposition.</u>
3 Top box: Breaks downwards with great force
Bottom box: Strong backwash

8. UK weather and climate

1 (a) **B** South-west (b) Warm, moist air and frequent rainfall
2 Temperate maritime
3 Freeze-thaw weathering
4 (a) The breaking down and removal of material along the coast
(b) Coastal erosion can cause coastal recession, which is when the coastline moves <u>inland. Frequent storms cause erosion and damage coastal landforms like spits.</u> (c) Strong winds increase the eroding power of the waves, so cliffs and other landforms will be eroded more quickly by processes such as abrasion. Storms bring heavy rainfall, which will increase the rate of mass movement so that more material slips to the base of cliffs and can be eroded. Landforms such as sand dunes can be removed by strong winds blowing away sand particles and by the high-energy waves caused by stormy conditions.

9. Erosional landforms

1 **A** = Headland; **B** = Wave cut platform
2 Box 2: The action of the sea causes erosion – mostly abrasion and attrition. This weakens the cliff base and helps with the cliff collapse.
Box 3: The upper part of the cliff (over the wave cut notch) is no longer supported and will collapse into the sea.
Box 4: Backwash removes the collapsed cliff face and this forms a wave cut platform.
Box 5: The process repeats and the cliff retreats, which increases the size of the wave cut platform.
3 A cave is formed when waves erode weaknesses in a headland such as joints. Over time processes such as abrasion cause the back of the cave to be eroded away to leave an arch in the headland.

10. Depositional landforms

1 (a) Bar (b) A bar is a long ridge made up of coastal sediment such as pebbles. A bar stretches across a bay, so that a section of seawater is cut off and dammed to form a lagoon.
2 The type of sediment depends on the local geology. Due to erosion, nearby cliffs are eroded, or sediment is deposited by constructive waves. The shape of a beach profile varies. Low energy constructive waves produce a low-angle profile and high energy destructive waves produce a steep-angle profile.
3 Box 1: Spits form when there is a change <u>in the direction of the coast.</u> Box 2: Prevailing winds mean waves <u>lose energy as they reach the coast.</u> Box 3: Sediment is transported by longshore drift and <u>deposited to form a spit.</u> Box 4: The spit may be curved because <u>there may be more than one LSD direction.</u>

11. Human activity

1 (a) Coastal <u>recession</u> (b) The loss of people's homes and disruption to communication networks, such as roads and railway lines (c) Weight of buildings makes cliffs more vulnerable to mass movement. Buildings can increase water going into the soil, making it more saturated and increasing mass movement.

2 Industry can increase air, noise and visual pollution. Visual pollution caused by large factories such as steelworks will have a major impact on the coastal environment. However, industry can bring wealth and jobs to coastal areas and increase of awareness and investment in protecting coastal landscapes.

3 Coastal recession and flooding can destroy wildlife habitats such as sand dune environments. Sand dunes are a natural form of coastal protection but are easily eroded by destructive waves. Increased deposition further along the coast will also affect the environment as spits may form, creating new landforms and habitats such as salt marshes.

12. Coastal management

1 Advantage = Prevent the sea removing sand and other sediment by longshore drift; Disadvantage = Other areas of coastline become exposed to erosion because sediment is prevented from reaching them.

2 (a) **D** Rip rap
(b) This technique reduces wave energy because <u>the gaps between the large boulders allow water through. This reduces wave energy and therefore erosion.</u>
(c) **B** A sea wall
(d) Sea walls help to prevent erosion by reflecting wave energy. There are no gaps, so sea walls do not reduce energy, but they create a strong backwash.

3 The advantage of using beach replenishment is that erosion protection takes place without the need for hard engineering as natural coastal processes are maintained.
The disadvantage of using beach replenishment is that the beach sediment will be moved over time by longshore drift. This means that the beach replenishment has to be constantly renewed or combined with another coastal protection method such as groynes, increasing costs.

13. Holderness coast

1 (a) Between 1846 and 1994 the coastline receded by more than half a kilometre. <u>Since 1955, almost a quarter of a kilometre has been lost.</u>
(b) 1.78 metres a year

2 *Example* **The Holderness coast** The geology and the location of the Holderness coast are significant factors. Geology is probably the most important factor as much of the coast is formed of soft boulder clay, which slumps after heavy rainfall. The slumped clay is then easily eroded and transported away by wave action and longshore drift. The Holderness coastline location is also very important as it faces the North Sea and is therefore affected by waves with a strong fetch. These waves have travelled a considerable distance before reaching the coast. This increases wave energy and therefore more erosion occurs. Storms in the North Sea are common, especially during the winter, and consequently the Holderness coast is eroded rapidly by destructive waves. The geology and the location therefore work together to make this a distinctive coastline of low, rapidly eroding and receding unstable cliffs.

River landscapes
14. Physical processes 1

1 The breakdown of rocks in situ

2 (a) Biological (b) <u>Chemical</u>

3 Box 1: Acids <u>released by the breakdown of vegetation increase chemical weathering.</u> Box 2: Roots <u>widen joints so freeze-thaw occurs more rapidly.</u>

4 The upper sections of the river valley are affected by weathering processes such as freeze-thaw action. This will produce loosened rock boulders over time. This weathered and eroded material moves down the valley sides by mass movement, especially sliding when the loosened material moves down rapidly due to gravity.

15. Physical processes 2

1 (a) **W** = Suspension; **X** = Traction; **Y** = Solution; **Z** = Saltation
(b) **Y**

2 Transport is the way in which the river carries eroded material or load.

3 **D** Deposition

4 Abrasion and attrition both involve erosion. However, abrasion takes place when the river's load rubs along the riverbed and banks, <u>wearing them away over time. Attrition is different because it involves the sediment transported in the river bumping and hitting against each other. This means that the transported load gradually becomes more rounded, smoother and smaller.</u>

16. River valley changes

1 (a) How a river's gradient changes between its source and its mouth
(b) **A** Upper course
(c) **C** Lower course

2 In the upper course, the river channel is narrow with steep sides. In the middle course, the channel becomes wider and the sides become less steep, and in the lower course the channel becomes wider, has a flat bed and the sides become less steep.

3 The landscape of the upper section of a river is formed by erosion, <u>therefore landforms include waterfalls and interlocking spurs.</u> However, the lower section of a river valley <u>is formed by mainly depositional processes although some erosion is also involved to produce landforms such as flood plains and river meanders.</u>

4 *Example* The River Holford flows across a series of different sedimentary rocks. Its source is on the Quantock Hill where the resistant sandstone means that the river erodes vertically to form steep-sided valleys and a steep upper course profile. In the middle section, the Holford flows across softer clays, which can be eroded laterally to produce a wide flood plain with a gently dipping long profile. At Kilve, where the river flows into the sea, it flows across limestone which has been eroded by past glacial events, giving a gentle profile.

17. Weather and climate challenges

1 (a) **C** Honister Pass December 2015 (b) The discharge would increase rapidly as more water reached the rivers, and the discharge might exceed bank full conditions, causing flooding.

2 Box 1: Erosion rates will be higher with greater discharge so landforms like <u>river valleys will become wider and deeper.</u>
Box 2: Load transport will <u>increase due to increased velocity.</u>

3 Increasingly frequent storms cause rivers to exceed bank full conditions and flood. More frequent hot, dry weather makes topsoil impermeable, causing rapid surface run-off and a sudden increase in discharge, resulting in flooding.

18. Upper course landscape

1 (a) Interlocking spurs (b) In its upper course, the river erodes vertically. The fastest water flow in the river channel will also naturally swing from side to side, causing maximum erosion on the outside of each bend. This will result in lateral as well as vertical erosion, especially if the local geology has lines of weakness such as joints. Over time, a series of spurs are formed, which are ridges of upland sloping down to the stream on either side of the valley.

2 (a) Gorges are formed when erosional process such as abrasion and hydraulic action cause a waterfall to retreat. The processes of undercutting and collapse are repeated over a long period of time, and the waterfall retreats, forming a steep-sided gorge.
(b) Waterfalls often occur where rocks of different hardness, and therefore resistance to erosion, occur together. The harder <u>rock (the cap rock) overlies the softer rock. The softer rock is eroded by hydraulic action and abrasion, undercutting the cap rock above. Over time, the cap rock collapses due to gravity. The height of the waterfall is increased with the formation of a plunge pool caused by hydraulic action and by fragments of cap rock and high-energy currents resulting in abrasion.</u>

19. Lower course landscape 1

1 Box 1: The river in normal flow conditions does not <u>exceed bank full.</u> Box 2: <u>During a flood there is an increase in discharge and the river exceeds bank full conditions.</u>
Box 3: <u>Floodwater on the floodplain loses velocity and deposits sediment.</u> Box 4: Repeated flooding means that <u>sediment builds up to form embankments called levées.</u>

2 **A** Point bar **B** Floodplain

3 Floodplains are formed by meanders eroding laterally as they travel downstream. Sediment is deposited on the opposite bank to form a point bar. The width of the floodplain is therefore determined by the amount of meander migration that occurs. When the river exceeds bank full and floods, it deposits layers of alluvium. These build up gradually over time to create the layers of sediment found on the floodplain.

20. Lower course landscape 2
1 (a) Lateral (b) (i) Width increases (ii) Depth increases
2 **C** Meander
3 *Example* When a river flows across its floodplain, it meanders. Deposition occurs on the inside of each bend where the current velocity is slower. At the same time, lateral erosion and undercutting occur on the outside of each meander where velocity is higher. Continuous deposition and erosion working together cause the formation of a pronounced meander with two outside banks becoming closer together. The narrow neck of land between the two meanders is finally eroded, either by lateral erosion or by the strong currents during flooding. A new, straighter river channel is formed and the abandoned meander loop forms a cut-off. Continued deposition seals off the cut-off from the river channel to form an oxbow lake. Erosion and deposition are therefore interdependent in all stages of the formation of oxbow lakes.

21. Human impact
1 *Example* Ploughing up and down a slope means that loose soil is transported into a river channel and can increase the amount of sediment deposited.
2 **C** By abstracting water for manufacturing
3 **A** Erosion decreases
4 Urbanisation has caused towns and cities to grow. Buildings are often constructed on river floodplains, which means that impermeable surfaces such as pavements are made and surface run-off increases. Water flows into the rivers increasing discharge. Rivers may be channelised, preventing erosion and deposition.
5 Urbanisation means that there are fewer permeable surfaces. Water flows into the rivers increasing discharge and therefore the risk of flooding. Field drains move water rapidly into streams and rivers, also increasing discharge and flood risk. Ploughing fields up and down a slope can increase the amount of sediment in rivers, reducing channel capacity and increasing the flood risk.

22. Causes and effects of flooding
1 (a) Labels correctly placed on graph.

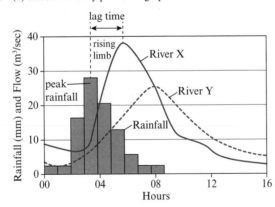

(b) The graph for River X has a much higher peak, rising to 38 m³/sec, but the peak for River Y is lower at 26 m³/sec. River X has a much steeper rising limb than River Y.
(c) The difference between peak rainfall and peak river discharge.
2 (a) River B has a much higher density of tributaries. This means that water resulting from intense rainfall events is taken to the main river very quickly, increasing discharge. The river may exceed bank full conditions and overflow onto the flood plain, causing flooding.
(b) Sudden snowmelt

23. River management
1 (a) Using man-made structures to control natural processes.
(b) Advantage = Long-lasting; Disadvantage = Expensive to construct
2 (a) Floodplain zoning
(b) Creating washlands means that parts of the floodplain are flooded frequently, which causes deposition and waterlogging. Agricultural land can therefore be turned into marshland. Floodplain zoning preserves the natural floodplain near to the river, but might mean that development on other parts of the floodplain is of higher density, changing the appearance and natural processes such as infiltration.

24. River Dee, Wales
1 **C** An area of upland glaciation
2 Box 1: Near the source, annual precipitation is very high. Box 2: The River Dee erodes the glaciated upland landscape formed of igneous and metamorphic rocks. Box 3: The middle course of the river is made of softer sedimentary rocks. The river both erodes and deposits sediment to form a floodplain. Box 4: More sediment is deposited in the estuary as velocity decreases.
3 *Example* The source of the River Dee is Dduallt, Snowdonia, a glaciated upland area consisting of metamorphic and igneous rocks. Precipitation exceeds 3000 mm per annum, so the Dee erodes vertically to create a steep-sided valley in its upper course. Near Bala, the river is controlled by the Dee Regulation Scheme, which means that human processes become more significant than physical processes in the river landscape. Reservoirs such as Lake Bala and Llyn Celyn prevent erosion in this part of the upper course, so erosion is replaced by deposition. The river is less regulated north of Chester and meanders across a distinctive wide floodplain formed by erosion and deposition. The channelised section to the west of Chester prevents erosion and deposition, forming a man-made landscape. As the Dee flows into its estuary, its velocity decreases and deposition of sediment transported along the channelised section causes extensive mudflats to form. The landscape of the Dee therefore varies depending on whether physical or human processes are more dominant.

Glaciated upland landscapes
25. Glacial processes
1 Freeze-thaw
2 Abrasion occurs when angular rocks become embedded in the base of the glacier. As the glacier moves these wear away the bedrock underneath by abrasion. However, plucking occurs when blocks of bedrock, loosed by freeze-thaw weathering, freeze to the base of the glacier and are pulled with the glacier as it moves out of a corrie.
3 Box 1: Till deposits are material deposited directly by ice. Box 2: Fluvioglacial material is glacial sediments deposited by meltwater streams.
4 The last major UK glacial period was the Pleistocene, about 12 000 years ago. Temperatures fluctuated above and below freezing causing freeze-thaw weathering, which caused glacial erosion. The ice accumulation exceeded ablation in winter so glaciers advanced. As temperatures increased at the end of the Pleistocene, ablation exceeded accumulation, so glaciers retreated.

26. Erosion landforms 1
1 A = Arête; B = Corrie; C = Tarn
2 Ice accumulates in a hollow on a mountainside. As temperatures fluctuate around freezing point, freeze-thaw weathering takes place. The accumulating ice deepens the hollow by abrasion as it flows out of the corrie and down the mountainside. The back wall of the corrie is made steeper due to plucking. There is maximum abrasion in the centre of the corrie where there is the largest mass of ice, resulting in a rock lip where the glacier flows out of the corrie causing less erosion.
3 (a) Roche moutonnée
(b) (i) Erosion by abrasion (creates a smooth stoss end)
(ii) Plucking, removing loosened rock (creates a rough lee side)

27. Erosion landforms 2
1 Two sets of contour lines parallel and close together indicate steep slopes. The two sets are separated by a linear area without contour lines, indicating flat land.
2 (a) The main glacier erodes a deeper glacial trough. A tributary glacier following a tributary river valley erodes a shallower trough, which can be seen in the upper part of the steep sides of the main trough.
(b) Glaciers usually move down a pre-existing river valley. The material transported by the glacier erodes interlocking spurs due to abrasion. Plucking also takes place so that the lower section of the interlocking spurs is planed off. This leaves a straight-sided glacial trough with truncated spurs.

28. Transport and deposition landforms
1 (a) **D** Terminal
(b) A glacier transports weathered and eroded material called till. When the glacier melts this material is deposited on the valley floor, usually as a series of ridges.

2 Box 1: The glacier is forced to flow over and around a band of resistant rock. Box 2: Erosional processes – abrasion and plucking – cause the resistant rock to steepen and become jagged, creating the crag. Box 3: The less resistant rock on the lee side is eroded to give a gentle gradient and moraine may be deposited.

3 Drumlins can be recognised by contour patterns. The contour lines form a concentric oval shape with one blunt end and one more pointed end.

29. Human activity

1 Upland glaciated landscapes tend to receive high rainfall. A number of glaciated troughs have been flooded to create reservoirs. The construction of dams changes the open appearance of a glacial landscape, and destroys natural habitats. The flooding of the area behind the dam also drastically alters the appearance and ecology of the glacial trough.

2 An advantage of tourism is it increases jobs for local people. However, a disadvantage is that footpath erosion can take place and paved paths are constructed, changing the appearance of the landscape.

3 Hill farming in glaciated uplands has created field boundaries (usually walls) and encouraged the removal of trees. Both change the appearance of the landscape. In some upland areas extensive conifer forests have been planted. This can prevent soil erosion but can also unbalance ecosystems, forcing some species out as the naturally occurring deciduous trees and shrubs are removed.

30. Glacial development

1 **D** Igneous, metamorphic and sedimentary

2 **A** In the valleys

3 During the last main ice advance, 18 000 years ago, there was an ice cap in the Snowdonia region. Weathering and erosion both by ice cap and smaller glaciers produced a distinct glaciated landscape. As the glaciers melted, glacial deposition took place in lower areas such as glacial troughs.

4 *Example* **The Snowdonia National Park** Processes such as weathering and erosion have continued in Snowdonia since the last major glacial period, but the most significant factor causing change is human activity. In the past, extensive areas of deciduous wood and shrubs were cleared for grazing sheep, changing the appearance and ecosystems of the region. Large areas of coniferous forest were planted more recently, again altering the appearance of the area. Slate quarrying was once an important industry in Snowdonia, but changed the natural landscape. For example, Dinorwic quarry, once one of Snowdonia's biggest slate quarries, has left waste slate tips and terraces, creating significant scars on the landscape. Walking is popular in Snowdonia, which has led to soil erosion of footpaths, increasing surface run-off and meaning that paved paths have to be built that conflict with the natural landscape. More than 360 000 people walk up Snowdon each year, so a railway and a visitor centre have been built, affecting the natural appearance of this glaciated mountain. Snowdonia can therefore be considered a largely man-made landscape.

Weather and climate
31. Global atmospheric circulation

1 (a) **C** Hadley cell
(b) The highest amount of solar radiation is received at the Equator. This causes warm air to rise to 15 km, causing low pressure at the Earth's surface. The rising air divides, cools and moves north and south to form two Hadley cells, one in each hemisphere. The air cools as it moves away from the heat source and becomes denser. Cooled air sinks at 30° north and south of the Equator, leading to high pressure. The air then returns as surface trade winds to the Equator due to pressure differences on the Earth's surface.

2 Solar radiation means air at the Equator is transferred north and south by the Hadley cells. Warm air rises because it is less dense. The air cools once it is away from the heat source, and sinks at 30° north and south of the Equator. Some of the cooled air moves back towards the Equator. The remaining warmer air travels towards the Poles, forming part of the Ferrel cells. At 60° north and south, the warmer air of the Ferrel cells meets colder polar air. The warmer air rises to form Polar cells. This air travels to the Poles, completing the heat transfer.

3 (a) **B** North Atlantic Drift
(b) Ocean currents transfer heat energy from areas of surplus (Equator) to areas of deficit (Poles). Wind-driven surface currents and deeper ocean currents move warm water towards the Poles and colder water toward the Equator. In the Arctic and Antarctic, water gets very cold and dense, so it sinks. Warmer water from the Equator replaces this surface water, creating ocean currents, such as the Gulf Stream. Cooled water flows back towards the Equator, forming cold currents, such as the Humboldt Current.

32. Natural climate change

1 Interglacial

2 Around 50 million years ago

3 (a) The Earth's orbit around the Sun changes shape approximately every 100 000 years. As a result, sometimes the orbit is more circular, which makes the climate slightly warmer (interglacial periods), and sometimes the orbit is more elliptical, which makes the Earth slightly cooler (glacial periods).
(b) Axial tilt cycle

4 (i) Historical sources, such as diaries. For example, accounts of the 1608 Frost Fair record the Thames freezing over and that Britain (and the entire Northern Hemisphere) was in a cool period known as the 'Little Ice Age'. (ii) Preserved pollen. This is evidence about warm and cold growing conditions. Each grain has its own unique shape, and walls made of assporopollenin, which is very chemically stable and strong. Therefore scientists use their knowledge about modern and historical distributions of plants to work out past climates.

33. Human activity

1 (a) Human activity causing levels of CO_2 and other greenhouse gases to increase in the atmosphere.
(b) The natural greenhouse diagram shows more heat energy radiated back into space. In the enhanced greenhouse gas diagram only about a quarter of the natural amount is shown being radiated back into space. The natural greenhouse diagram shows significantly lower amounts of CO_2, CH_4 and N_2O in the atmosphere than the enhanced diagram.

2 (a) **D** Photosynthesis
(b) The greater use of transport due to increased car ownership and flights is releasing more greenhouse gases into the atmosphere and increasing the greenhouse effect.

3 Climate change in areas near the Equator, such as Africa's Sahel, is causing less predictable rainfall and longer periods of drought. This is lowering crop yields and leading to food shortages and malnutrition. Melting ice sheets and retreating glaciers add water to oceans, causing global sea levels to rise. Many low-lying islands, such as the Maldives, face greater flood risk from rising sea levels. There will be more coastal flooding, loss of beaches and loss of coral reefs. This will have negative impacts on the economy, which depends to a large extent on tourism. Some islands will have to be evacuated, causing a breakdown in social structure and loss of employment. The fragile coral reef ecosystem will be destroyed because corals cannot grow in deep water, resulting in a reduction in biodiversity.

34. The UK's climate

1 (a) (i) Medieval Warm Period (ii) Little Ice Age
(b) *Example* During the Little Ice Age (1600–1685) temperatures were low enough to freeze the Thames, due to increased volcanic activity that caused a major ash cloud and decreased solar radiation.

2 (a) 6 °C
(b) Both the temperature and precipitation graphs show a summer maximum. The highest temperature (17 °C) is in July/August and coincides with the highest precipitation (60 mm). The lowest temperature is 3.5 °C in February, which is also when the lowest precipitation of 25 mm occurs.

3 The UK is located between 50° N and 60° N. Therefore the UK is near the boundary between the northern Ferrel and Polar atmospheric circulation cells. This is where warmer air from the south and cooler air from the north meet, causing unsettled weather due to the formation of depressions over the Atlantic, which then travel east to the UK.
The North Atlantic Drift flows northwards near to the west coast of the UK. In the winter this makes the UK climate milder than would be expected for its latitude. The main or prevailing wind affecting the UK comes from the south-west. This air travels long distances over the Atlantic Ocean and over the North Atlantic Drift bringing moisture, leading to rainfall along the west coast of the UK. The prevailing wind also modifies temperatures so that the western UK is warmer than would be normal for its latitude. The UK's geographical position is therefore very significant and results in the UK having a warm maritime climate.

35. Tropical storms

1 Tropical cyclones are mostly located in the tropics between 23° north and south of the Equator. This is because these storms are powered by warm ocean temperatures – the water needs to be above 26.5 °C and this only occurs in late summer and autumn in the Tropics. Secondly, tropical storms form in areas where a rotation force, created by the Coriolis effect, forms part of the atmospheric global circulation. This causes the storms to develop and rotate.
2 GIS
3 Track
4 **D** Eye
5 Tropical cyclones need a source of warm, moist air. This is produced by tropical oceans with sea surface temperatures of about 27° C. Winds converge and cause air to rise and storm clouds to form. As the air rises, it accelerates and begins to spiral due to the Coriolis effect. This means an area of very low pressure forms at the centre of the converged storms.

36. Tropical cyclone hazards

1 (a) **A** Saffir-Simpson
  (b) (i) Wind speed (km per hour); (ii) Storm surge (metres)
  (c) **C** 3
2 **C** Landslides
3 (a) Cyclones produce winds as strong as 250 km/hour. These are caused by the flow of very warm, moist, rapidly rising air, leading to the development of a centre of low pressure, or depression, at the surface. As air moves from high to low pressure this creates strong winds.
  (b) Heavy persistent rainfall results from the rapidly rising moist air, which cools as it rises. The water vapour condenses to form water droplets and extensive storm clouds, such as cumulus-nimbus, followed by heavy rainfall.
  (c) Tropical cyclones can cause a large mass of water to reach coastal areas. Reduced pressure at the centre of the cyclone sucks seawater upwards, increasing sea levels and creating storm surges. Strong winds also pile up seawater and increase flooding.

37. Hurricane Sandy

1 19 + 21 / 2 = $20billion
2 Hurricane Sandy formed in ideal conditions in the Caribbean Sea. Warm water was available to strengthen the tropical storms, which grouped together to form the hurricane. After passing the Bahamas, the hurricane turned north-west and strengthened to a category 1. By the time the hurricane reached the USA coast, wind speeds of 129 km/h were recorded, causing severe damage. Storm surges, due to the extreme winds, caused most of the damage to east coast states, destroying homes and infrastructure.
3 Raw sewage entering rivers
4 Twitter
5 The worst social impact of Hurricane Sandy was the number of deaths. 150 people were killed, which caused immense trauma to families in the affected regions. 100 000 homes were destroyed and on Long Island a further 2000 were made uninhabitable, causing the break-up of communities. Many areas were left without electricity – even months after the hurricane there were still 8 200 people without power – again causing major social problems.

38. Typhoon Haiyan

1 (a) 5
  (b) North-west
  (c) **A** The Philippines
2 Mangrove trees were damaged and uprooted across the islands. This disrupted the fragile ecosystems such as coral reefs. By trapping nutrients and sediments from drainage, mangroves protect coral reefs, seagrass meadows and coastal waters.
3 *Example* Responses to tropical cyclones tend to be more effective in developed countries such as the USA than in developing countries such as the Philippines. The USA government was able to allocate billions of dollars to help support victims of Hurricane Sandy and to help rebuild the coastal areas affected. In the Philippines, which was affected by a tropical cyclone, Typhoon Haiyon, of similar magnitude, the government declared a 'state of national calamity'. However, aid for emergency shelters, water and food was largely supplied by the international community (for example, the UK provided £10 million).

The speed of response tends to be more rapid in developed countries, which have the technology in place to search for victims and provide emergency infrastructure. This technology is largely absent in developing countries due to a lack of finances and investment. This causes delays, which increases the risk of death and disease, and means that developed countries have more effective strategies for managing the impacts of severe weather events.

39. Drought causes and locations

1 Arid environments have a permanent low precipitation – usually between 10–250 mm a year – whereas drought conditions are caused by a temporary period of low precipitation. Both are associated with low pressure and little or no cloud cover.
2 (a) Meteorological
  (b) *Example* In the UK, high pressure called a blocking anticyclone forces low pressure systems that bring rain around it, either to the north or south of the UK. This means no rain falls over all or part of the UK for weeks, resulting in drought.
3 The meeting of Hadley and Ferrel circulation cells in the upper atmosphere causes descending dry air, so condensation does not occur. Clouds cannot form and there is little precipitation (for example, in Africa's Sahel).
4 (a) 7 cm (accept 6.5–7.5)
  (b) Since 1990, the overall trend is that rainfall for each year is less than 0 cm/month. The lowest amount was recorded in 2002, approximately −2 cm/month. There are some anomalies with positive figures: for example, 2010 with 2 cm/month.

40. California, USA

1 Droughts can cause subsidence as groundwater levels drop, causing land to settle at a lower level. This can damage the foundations of buildings, making them uninhabitable. Seawater can be drawn inland as pressure in groundwater supplies drops. This causes water supplies to become saline and not suitable for drinking.
2 A large blue-green algal bloom has developed in the Sacramento-San Joaquin Delta. This type of algae produces toxins that in high concentrations are lethal to fish and people. In forest areas, forest wildfires have become frequent as trees and other vegetation become dry, damaging animal and plant habitats and killing wildlife, therefore forest ecosystems are destroyed.
3 In order to save water, 542 000 acres have been taken out of crop production resulting in loss of income for farmers. Overall the drought is costing California US$2.7 billion a year, meaning less state money is available to spend on services for people, such as education.
4 The general trend of reservoir storage for each year is that the highest storage occurs in May and June and the lowest storage in November and December. The highest storage overall was in 2011. All other years show a decrease in storage, and the levels in 2015 are almost half those of 2011.

41. Ethiopia

1 The overall pattern for food aid is an increase on the 2001 figures in 2007 but a decrease to the lowest level in 2013. In 2007, $150 million was given, but by 2013 this had fallen to $125 million.
2 The 2015 drought was the most severe for 30 years. Prolonged lack of rainfall meant that important ecological areas such as the Borkena Wetland became dry, resulting in the loss of habitats for fish and wildlife. Vegetation in grassland regions died, causing the near extinction of Grevy zebras and forest fires have led to the loss of 200 000 hectares of forest every year, damaging ecosystems.
3 *Example* In the USA, California has had drought conditions for several years. As a developed country, the USA has the finances for the government to organise public education programmes such as Save our Water, and the local state government has responded by passing laws which require a 25 per cent cut in water consumption. In contrast, Ethiopia, a developing country where drought can result in death and malnutrition, relies heavily on overseas aid to provide food. In 2015, the USA government gave US$128.4 million in aid. The University of California is funding long-term research projects on the effective management of groundwater. However, a lack of technical support and funding in Ethiopia means that organisations such as Oxfam and WaterAid are helping local communities with small-scale, appropriate technology

such as wells, which are short- to medium-term solutions to the problem. Responses to drought therefore reflect the country's income and level of economic development.

Ecosystems

42. The world's ecosystems
1 (a) (i) Low temperatures (ii) Low precipitation
   (b) **D** Central South America
2 Temperatures for the tropical forest graph are above 26 °C all year and rise to a maximum of 27.5 °C in October. However, temperatures in the boreal forest only reach above freezing between May and <u>October. Precipitation occurs all year for both ecosystems. The highest precipitation for the tropical rainforest is between November and May (maximum April 325 mm), while the maximum for the boreal forest is in August (75 mm).</u>
3 Growing seasons are longer in warmer locations near the Equator and tropical rain forests are located in equatorial areas and boreal forests at cooler higher latitudes.
4 Altitude

43. Importance of the biosphere
1 The biosphere
2 (i) Timber (ii) Straw
3 One fuel resource from the biosphere is bioethanol. This is made by converting plant products such as sugar cane into fuel by a series of processes.
4 One-quarter of all prescription drugs come directly from, or are derivatives of, plants. <u>For example, periwinkles are used to treat leukaemia and Hodgkin's disease. This means that the tropical rainforests are important medicinally. Plants are also important for providing painkillers. For example, poppy seeds are used to make morphine.</u>
5 (a) 4% (accept 3–5%).
   (b) Increased water consumption means there is less water in the biosphere for other areas (for example, wetlands) that then might dry out, causing loss of habitats and species.

44. The UK's main ecosystems
1 (i) <u>Consist of peat bogs and rough grassland</u> (ii) Found in upland areas
2 **B** Waterlogged soils; **F** Low in nutrients
3 (a)

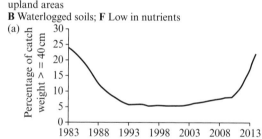

   (b) Overfishing fish species such as cod.
   (c) Fertilisers used by farmers can lead to eutrophication due to chemicals reaching the sea. This causes a surplus of nutrients, leading to a dense growth of plants, which reduces the amount of oxygen in the water and causes the death of fish and other organisms. The development of coastlines for industry can lead to destruction of plant and wildlife habitats such as salt marshes and the degradation of other sensitive habitats.

45. Tropical rainforest features
1 (a) **A** Biomass **B** Litter
   (b) High annual temperatures, year-round rainfall and a continuous growing season all mean that there is a dense, rapidly growing vegetation with a high biomass.
   (c) (i) The nutrients in the soil are rapidly taken up by plants as nutrients are needed for the high rates of growth and photosynthesis. (ii) <u>High levels of precipitation wash any remaining nutrients down to the lower sub-soil level. This results in leaching.</u>
2 **B** Plants
3 (a) Chemical weathering
   (b) The warm moist conditions due to the high levels of precipitation and solar radiation mean that rapid chemical reactions occur causing the breakdown of rocks.
4 Indigenous tribes spread seeds of plants when eating fruit and nuts. This helps trees and other plants to spread to new areas, maintaining the biodiversity of the tropical rainforest.

46. TRF biodiversity and adaptations
1 **B** Large percentage of body fat
2 **D** Deciduous

3 Box 1: Emergent trees <u>grow above the canopy to obtain the maximum sunlight for photosynthesis.</u> Box 2: Lianas are woody vines that climb up tree trunks so that <u>they reach sunlight for photosynthesis.</u> Box 3: Many plants have modified leaves with drip-tips, which <u>allows rainwater to run off and prevents the leaves rotting.</u> Box 4: The tallest trees often have buttress roots, which <u>help to support and make the trees stable.</u>
4 The number of varieties of plants and animals in a particular habitat or ecosystem.
5 Tropical rainforests have high levels of biodiversity because they are the oldest global ecosystem, so a wide range of species has evolved over a long time. Their location, between the Tropic of Capricorn and the Tropic of Cancer, means long hours of sunlight and warm temperatures, which are excellent for photosynthesis. Consequently food is always available for insects and animals, which leads to biodiversity.

47. TRF goods and services
1 **B** Home to indigenous tribes
2 (i) Food, such as fruits and nuts (ii) Timber for manufacturing furniture, for example
3 (a) Figure completed with correctly sized bars, using data from table.
   (b) Climate change (e.g. less rainfall) threatens the survival of plant and animal species, which are specifically adapted to the high rainfall conditions present, and this reduces biodiversity. Drier conditions could stop 'cloud functioning' and reduce the water available for plant growth, which would have a negative impact on biodiversity.
4 Climate change may result in reduced precipitation and long periods of drier conditions. This will <u>greatly reduce the biomass store. Therefore the layered structure of the rainforest will be reduced because there will be fewer nutrients available for the canopy and other sections of the forest. Drier conditions make forests more vulnerable to fires, which destroy the structure of the forest.</u>

48. Deforestation in tropical rainforests
1 Global demand means that mining occurs. Extensive deforestation takes place near the mining area and along the routes built to remove ores.
2 The overall trend for the resources graph is downwards over time but the trend for the population graph is upwards. Both graphs show a levelling out after 2000. The population graph reaches a peak after 2000 and then declines, whereas the resources graph shows the largest fall before 2000.
3 (a) **C** 90
   (b) Deforestation is taking place for large-scale agriculture such as oil palm plantations. There is a high global demand for palm oil and therefore <u>farmers can increase their income by clearing forest and planting oil palm plantations.</u>
4 Rapid population growth.

49. Tropical rainforest management
1 Using resources in a way that will benefit current generations and future generations.
2 (a) This is responsible tourism to unspoilt areas that preserves <u>the environment and improves the wellbeing of local people.</u>
   (b) Ecotourism provides an alternative form of employment for local people that doesn't involve deforestation: for example, by creating jobs such as guides for the Manu Cloud Forest Canopy Walkway.
3 Timber firms are using reduced-impact logging, which ensures that only certain trees are felled, as this can be more profitable than conventional methods of timber extraction.
4 Encouraging conventional logging
5 Only selected mature trees are cut down, while 'seed' trees are left to grow. This leads to quicker regrowth in the gaps left by the felled trees. The direction of falling trees is calculated to reduce damage to other trees. This results in less fragmentation and more rapid regeneration of the rainforest than conventional felling.

50. Deciduous woodlands features
1 **B** The UK; **C** Japan; **D** China
2 (a) (i) Biotic = Trees (ii) Abiotic = Soil
   (b) Box 1: There is more light and water available in the <u>spaces between tall trees for the sub-canopy layer.</u> Box 2: The herb layer is formed of plants such as bluebells, which <u>flower</u>

early before the light is blocked out. Box 3: Plants such as mosses grow on the ground layer because <u>they are adapted to wet and shady conditions.</u> Box 4: The brown earth soil is fertile because <u>of a thick layer of leaf fall each year.</u>

3　**B** Precipitation all year with a winter maximum; **C** An annual temperature range of approximately 12 °C

4　The soil store results from a large input of nutrients each year due to autumn leaf fall and a second input from weathered rocks. Precipitation in deciduous woodland regions is not high enough to leach the nutrients, so a large store accumulates. Uptake by plants is mostly restricted to the period between April and September, and the vegetation is restricted to one main layer, so the biomass store does not become larger than the soil store.

51. Deciduous woodlands adaptations

1　(a) 4700 (kilocalories/square metre/year)
　　(b) Deciduous woodlands have a canopy of close-growing trees so there is extensive leaf cover to carry out photosynthesis. The ecosystem also has four layers, which increases productivity.

2　Some animals, such as hedgehogs, hibernate in winter. This means that they do not have to search for food such as small insects, which become very scarce during cold winters, and reawaken in the spring when food becomes available. Squirrels have adapted by storing food such as acorns, burying it in spring and summer to use in winter.

3　Box 1: Deciduous leaves are <u>broad and thin</u>. This enables them to carry out maximum photosynthesis <u>as they have a large surface area.</u> Box 2: Deciduous trees drop their leaves in autumn because <u>this helps to reduce the rate of transpiration and conserves water.</u>

4　For stability.

52. Deciduous woodlands goods and services

1　Deciduous woodlands act as carbon stores. It is estimated that UK woodlands take in 1 million tonnes of carbon per year.

2　**B** Fuel; **D** Timber

3　Climate change may result in drier summers. Trees affected by drought may die, which would lower biodiversity as insects and <u>animals would not have a food supply</u>. Biodiversity could also be reduced because <u>the warmer conditions might cause a rise in diseases that could affect vulnerable species, again causing a loss in biodiversity.</u>

4　(a) It is becoming earlier
　　(b) This is due to climate change. Winters in deciduous woodlands are becoming milder with less frost and snow but more rainfall. The warm, wet conditions in spring are encouraging trees to come out of dormancy earlier. Therefore leaves are now appearing at the beginning of April rather than at the end of April as they did in the 1950s.

53. Deforestation in deciduous woodlands

1　(a) Figure completed with correctly sized bars.
　　(b) Population growth
　　(c) Population growth increases the demand for homes. Land is required, which puts pressure on greenbelt sites such as deciduous woodlands where houses can fetch higher prices, leading to woodland clearance.

2　**A** More farmland needed; **C** Increased use of pesticides and herbicides

3　A disadvantage is that coniferous trees support lower levels of biodiversity as they reduce the amount of light and precipitation from reaching the woodland floor.

54. Deciduous woodlands management

1　*Example* Areas that contain extensive deciduous woodlands, such as the New Forest, tend to be popular tourist attractions. Visitors cause increased litter, disturb wildlife, erode footpaths and increase air pollution. In addition, softwood and hardwood timber is extracted for commercial use and 40 per cent of the woodland is privately owned, and is often left unmanaged. Sustainable management is therefore needed to regulate these activities and to ensure that the woodland is conserved for future generations.

2　**B** Tourism

3　Careful management, with dedicated walking and cycling routes in more fragile areas, means that <u>footpaths and ecologically fragile areas are not eroded.</u> Landowners are given grants to plant native tree species, which means <u>that the biodiversity of the area is preserved as deciduous woodlands have a higher biodiversity than coniferous woodlands.</u>

4　*Example* Tropical rainforests such as the Amazon rainforest are usually located in a developing or emerging country. The country often relies on the income from logging and timber extracting, which forms a significant part of the country's economy. In 2015, the value of timber-related exports was $3 billion. Regions with deciduous woodlands, such as the New Forest, UK, are more often developed, therefore the income from timber extraction forms a small part of the country's budget. Consequently it is easier to regulate timber extraction from deciduous woodlands.
Much of the management of deciduous woodlands in the UK is controlled by strict laws and funded by the government: for example, landowners are paid grants to plant native tree species in the New Forest. All logging in the Amazon area requires a permit and a formal management plan. These are issued by the Environmental Institute of Brazil. However, despite this requirement not all logging is carried out legally. Consequently it is much more difficult to make effective management plans for tropical rainforests as the value of the timber is much higher and it is more challenging to enforce laws protecting the area.

Extended writing questions
55. Paper 1

Economic factors
- Land has been cleared for cattle ranching due to increasing demands for beef. For each kg of beef produced, 37 km² of rainforest is destroyed.
- In the past 20 years, Costa Rica has lost much of its rainforest to beef cattle ranching.
- 50 per cent of rainforest destruction is due to cattle ranching.
- Extensive areas are cleared for plantations, particularly palm oil. It is estimated that 98 per cent of the Indonesian rainforest will be cleared for palm oil production by 2022.
- Mineral resources, such as iron ore (Brazil's main export), come from rainforest areas.
- Logging involves cutting down mahogany and teak trees. About 3 per cent of the Amazon rainforest is lost every year due to logging.
- Creating infrastructure, such as building roads to access and remove products, involves felling trees.

Social factors
- Governments and international aid agencies have encouraged people to move into rainforests.
- The Transmigrasi Program in Indonesia caused an average annual loss of 200 000 hectares.
- Population pressure.

Assessment Objective 3
- Economic and social factors both have the same result: the loss of extensive areas of rainforest and its associated biodiversity.
- However, the direct economic causes (particularly felling for cattle ranching and the establishment of plantation agriculture) are causing more deforestation, fragmentation and loss of habitats than social causes.
- Some of the social causes are directly linked to the economic factors. The Transmigrasi Program encouraged people to move into rainforests to carry out small-scale farming for economic reasons, such as reducing poverty in urban areas.
- In the long term, economic factors may be important in helping to reduce the rate of deforestation as incentives such as ecotourism become more important.

COMPONENT 2: THE HUMAN ENVIRONMENT
Changing cities

56. An urban world

1　The increase in the proportion of people living in built-up areas instead of rural areas.

2　(a) Both the graphs show an increase in the number of people living in urban areas. However, the graph for developed regions increases much more slowly than the developing graph, rising from 400 million in 1950 to a projected figure of 1100 million in 2050. The developing regions graph <u>rises rapidly from 500 million in 1950 to over a projected 5000 million in 2050. Therefore the projected population for developing regions is five times as large as the projected developed population.</u>
　　(b) Factory jobs in urban areas during the industrial revolution meant people moved from rural to urban areas.

3　Shanty towns.

4　In developing countries, illegal, unplanned shanty towns often develop as there is limited affordable accommodation for the people moving to the urban areas. The increase in population means that air, noise and water pollution become problems: for example, few shanty towns have proper sewage or drainage systems so raw sewage flows into rivers.

57. UK urbanisation differences

1. (a) Box 1: Scotland has a relatively low population density because upland relief makes transport and construction expensive. Box 2: There are fewer large urban areas in the north of England due to upland areas, e.g. the Pennines, where it is difficult to build settlements. Box 3: London is the capital and main finance centre of the UK, so the population density is the highest in the UK (5100 per km²). Box 4: Ports such as Liverpool have a high population density as they provide employment in industries.
   (b) **D** 266 people per km².
2. **A** Birmingham; **C** Leeds
3. The degree of urbanisation is related to the amount of employment available. In areas with a high level of employment, such as London or Birmingham, large-scale conurbations have developed where several former towns have been joined together. Improved infrastructure affects the degree of urbanisation. Areas accessible by motorways or fast rail links are more desirable and grow rapidly as commuting and the movement of goods becomes more economic and rapid.

58. Context and structure

1. *Example* Birmingham has excellent road transport links because major motorways, including the M42/40 from London and the M1 which runs north to Stoke-on-Trent and Liverpool, interconnect in Birmingham. Birmingham is also the centre of a major rail network as the newly developed New Street station links Birmingham with Birmingham airport, London, Manchester, the Midlands and north Wales.
2. *Site of a settlement* describes the physical nature of where it is located.
3. A settlement's location in relation to surrounding human and physical features
4. *Example* The CBD or Central Business District of Birmingham is situated roughly in the centre of the city where the highest-value land is located. This main financial centre of a city and the offices, national chain shops, restaurants, expensive apartments and hotels can afford the rates charged. The inner city – for example, Sparkbrook – consists of tightly packed lower-quality housing while the suburbs and the edge of Birmingham consist of lower-density housing as land prices decrease with distance from the CBD and there is more land available for development.

59. A changing UK city

1. (a) Suburbanisation.
   (b) Counter-urbanisation is the movement of a large number of people from urban areas to the surrounding rural areas.
   (c) People move from urban areas because traffic congestion and emissions leads to poor air quality, so they are attracted by the better air quality in rural areas.
2. **D** People move back to the centre of urban areas
3. (a) Map completed using correctly sized bars.
   (b) *Example* The population pyramid for Birmingham shows that there are the highest number people in the age group 20–24. This is because young people from other countries migrate to Birmingham for work.

60. Globalisation and economic change

1. Map completed using correct colours from key.
2. *Example* Population growth in Birmingham has occurred because of an increase in the birth rate, which was 2.08 children per woman in 2013, above the average of 1.85 for England and Wales. Increased international migration has resulted in a larger population. Between 2012 and 2013 this resulted in an estimated increase of 3400.
3. The NEC (National Exhibition Centre) was built near junction 6 of the M42 motorway, east Birmingham, to encourage major show events to move away from London as part of government decentralisation policy.
4. A reduction in the size of the manufacturing sector.
5. *Example* Deindustrialisation has occurred in Birmingham due to the national shift from a secondary to a tertiary and quaternary economy. During the recession of the 1980s, a number of manufacturing companies were forced to reduce their production, such as the van maker LDV, which made 850 workers redundant and the engineering group GKN, which cut 564 jobs. The increase in unemployment led to schemes including Support for Deprived Neighbourhoods,

designed to increase social housing, and the Birmingham Big City Plan, which was implemented to revitalise Birmingham and create a more appealing city centre and increase employment, training and leisure opportunities for future generations. However, the loss of MG car manufacturing at Longbridge in 2005 meant that 6000 people lost their jobs and unemployment almost doubled in the wards near the car factory. However, Birmingham has since been transformed and diversified into a major retail centre, with the Bullring and the Grand Central centres changing the dependence on manufacturing to retail, high-tech companies and banking.

61. City inequalities

1. The areas in the most deprived 5% form a discontinuous band from west to east across the centre of Birmingham. There are fewer areas in the very centre and the far west. There are no areas in the north and north-east of Birmingham, and only a few small areas scattered in the south-east and south-west.
2. *Example* One important reason for increased inequality in Birmingham is the decline in manufacturing industry, such as car manufacturing at Longbridge which affected people living in the inner city.
3. *Example* In Birmingham, the natural increase in population and immigration is causing an increase in the population of inner city areas such as Sparkbrook. This results in pressure on health and education services, which lack both finance and specialists. In the suburbs, such as Harbourne, population density is more constant, and therefore the provision of services is higher per head of population.

62. Retailing changes

1. Online shopping is convenient, because it can be carried out from home or from a mobile phone. It also allows people to carry out 24-hour shopping, so people who do not have time to go shopping during normal business hours can still do food and grocery shopping.
2. **D** Retail and offices
3. (a) Selling goods directly to consumers
   (b) Box 1: Merry Hill edge-of-town shopping centre is about 30 km west of Birmingham CBD. Edge-of-town shopping centres have developed due to demand for free and available parking. Box 2: The Grand Central shopping centre forms part of the development to regenerate Birmingham's centre by attracting investment and providing jobs. Box 3: The Bullring was redeveloped because out-of-town shopping centres were attracting customers, resulting in the decline of the CBD.

63. City living

1. (a) Conversion of waste into reusable material.
   (b) 3.4%
   (c) Recycling reduces the need to send waste to landfill sites. This is important because it helps to prevent greenhouse gases (especially methane, which is 21 times more harmful than CO_2) being released into the atmosphere.
2. Affordable housing is owned and let by housing associations, but higher rents are charged than those for social housing. The money from rents is used towards building more houses. However, energy-efficient housing can be housing owned by anyone who uses less energy such as electricity to provide the same level of energy. Affordable housing can be energy-efficient.
3. Hybrid buses that have electric battery packs and a diesel engine produce 40 per cent less carbon dioxide (CO_2) emissions than traditional buses, so reduce emissions in urban areas.

64. Context and structure

Example
1. (a) Mexico City's original site was an island in Lake Texcoco with an altitude of 2200 metres. Lake Texcoco was drained as the settlement expanded to provide a very flat site with saturated clays underneath.
   (b) Mexico City is situated in the Central Plateau, a flat landscape surrounded by mountains and volcanoes. Mountain ranges of the Sierra de Guadalupe are to the north and the Sierra de Ajusco are to the south.
2. Box 1: International routes such as Federal Highway 57 are important because they link to the USA for trade. Box 2: The International airport links Mexico City with South America, the USA and Europe. Therefore this creates an economic and transport hub, encouraging development. Box 3: Motorways connect to industrial towns such as Toluca, which means that goods can be transported to the main market.
3. *Example* The CBD of Mexico City includes the Santa Fe area, one of the major business districts, with modern-built office

blocks, the financial hub including the Mexican Stock Exchange and government buildings. There are high-value retailers and shopping malls, especially in the newly developed Santa Fe area.

65. A rapidly growing city

1 (a)
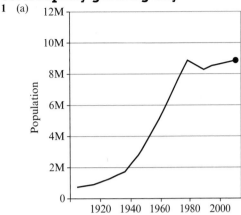

(b) (i) *Example* During the 1950s a high birth rate and a fall in the death rate meant the population of Mexico City grew rapidly. This rapid increase has continued with improved health care which lowers infant mortality.
(ii) Economic investment has **increased** job opportunities through the investment in the construction of factories and offices in the city, for example the Santa Fe area, leading to an increase in the population.

2 **A** People come to a country from another country

3 *Example* One of the most evident impacts of migration is the rapid increase in the city's population size. This causes demand for housing which causes more self-built housing at the city edge, creating large informal settlements. It is difficult to supply clean water and drainage, putting significant pressure on the city's reservoirs and underground aquifers. Nearly 30% of the homes in Mexico City do not have internal water supplies and some do not have access to fresh water at all. This increases the risk of disease and puts significant strain on health services, which struggle to cope with the increasing demands from the expanding squatter settlements. Many migrants moving to the city are young adults increasing the population under 30 but the lack of jobs is increasing the rates of crime. Not only is this adversely affecting law and order enforcement, but it is also affecting Mexico City socially as the fear of crime is leading to segregation of people, with the wealthier living in gated communities. This increases the gap between rich and poor and prevents social cohesion. Therefore the economic and social impacts of migration are both significant and problematic.

66. Increasing inequalities

1 Rapid internal migration, often rural to urban, plus natural increase causes a rapid increase in population. This creates housing shortages and squatter settlements develop to provide homes.

2 *Example* A rapid increase in population and urbanisation leads to an increase in cars and other vehicles. Emissions from vehicles (for example, 3.5 million cars in Mexico City) cause high levels of air pollution, especially greenhouse gases. Mexico City is unable to cope with the volume of waste, such as sewage, resulting in waste being dumped on the streets and in water supplies, causing pollution.

3 *Example* In Mexico City, 27% of homes do not have an internal water supply and 2% do not have access to water at all. This is because these homes are usually squatter settlements which are both unplanned and unregulated. They are often built on land such as rubbish tips which are not connected to infrastructure, such as water supplies. Informal squatter settlements grow so rapidly that the authorities do not have the time or finance to connect all the houses to clean water supplies.

4 Not having enough paid work

5 *Example* In Mexico City, the minimum wage should be $4 per day, but about 1 in 3 workers is paid less. In the CBD wages are much higher, with the top 20 per cent earning as much as 13 times more than the bottom 20 per cent. This increases the inequality between the rich and poor in the city. Variations in education provision also increase inequality as the poorest 10 per cent only attend school for two years while the richest 10 per cent average 12 years.

67. Solving city problems

1 *Example* In Mexico City, improving living conditions in squatter settlements often uses the bottom-up approach. These are often small-scale projects that have the advantage that they teach local people new skills such as bricklaying. As the bricks are made from the clay deposits around the city, the projects also have the advantage of being appropriate and sustainable in the long term.

2 *Example* Community projects include those such as Cultiva Ciudad, which is working with local schools to educate children on how to garden. Rooftop gardens have been established which provide residents with healthy food, helping to reduce food shortages and obesity. Other schemes include local people working in informal housing areas to raise money to help with building schools and health centres, which will help to improve the health and future prospects of residents.

3 *Example* Governments are introducing or expanding public transport systems to reduce air pollution. For example, in Mexico City the government introduced the Metrobus in 2005 to reduce 35 000 tonnes of CO$_2$ emissions each year. The rapid bus transport system moves approximately 250 000 people per day and has reduced journey times by 30 minutes which encourages people to use public transport and helps to reduce levels of pollutants like nitrogen dioxide.

4 *Example* In Mexico City the government set up a scheme in 2011 to help reduce the extensive waste problem. They introduced a trading system, which swaps trash for food. A 'barter market' was set up with residents exchanging waste for vouchers. These vouchers were then traded for food with local farmers, resulting not only in a reduction in waste but also an improvement in the diet of many residents.

Global development
68. Defining development

1 How advanced a country is compared to others. **Development** also refers to the standard of living and quality of life of its human inhabitants.

2 Development can take place by investment in farming. Improved seeds, fertilisers and mechanisation can improve yields, providing more food to eat and a surplus to sell for an increased income. Improving education will raise literacy rates and improve skills and job prospects, so the economy will be able to diversify into the secondary, tertiary and quaternary sectors, which will increase development.

3 Food security

4 (a) When every person has access to enough safe and affordable water.
(b) Most of the countries are north of the Equator in a band that stretches from Africa east to Pakistan and Uzbekistan. The countries are mainly scattered, only Egypt and the Sudan form a large continuous area. There are no countries at extreme risk in North America, South America, Europe or Oceania.

5 Food security is essential for development. GDP growth due to agriculture is approximately four times more effective in reducing poverty than other types of growth. It is estimated that a 1% increase in agricultural yields results in a 0.6–1.2% reduction in the number of people living below $1 per day, so this increases their income and their ability to become economically active.

69. Measuring development

1 **B** CBD

2 (a) 87 + 86/2 = 86.5
(b) The annual index of perceived corruption shows that high levels of political corruption lead to exploitation, inequality and poverty and therefore low levels of development.

3 The Gross Domestic Product is divided by the population to give GDP per capita, the higher the GDP per person, the higher the level of development. The HDI, the Human Development Index, includes Gross National Income, life expectancy and average years in education to produce an indicator of a country's development level. The higher the number, the higher the level of development.

4 Inequality measurements are often based on the distribution of income and economic inequality among people living in a country or area. One way of measuring this is to use the Gini coefficient. This is expressed as a ratio from 0 to 1.0 means everyone has the same income and there is no inequality; 1 means one person has all the income and there is complete inequality.

70. Patterns of development

1 (a) **A** A measure of development based on social and economic levels
   (b) Map completed using correct colours from key, i.e. UK = red; Brazil = orange; India = yellow/green.
   (c) Most of the highly developed countries are in the northern hemisphere, especially in northern America and Europe. Highly developed countries in the southern hemisphere tend to be dispersed, <u>such as Australia and Argentina. Overall, there are a higher number of countries that are highly developed in the northern hemisphere than the southern.</u>
2 Physical factors are important. Areas that are remote or are relatively inaccessible because of upland areas, such as north Wales, tend to have a lower level of development than highly accessible areas such as south-east England. Economic factors including employment rates and salaries are significant. The south-east of England has a high disposable income per person and an unemployment rate of 3.8%. However, in Scotland, where disposable incomes are lower, the unemployment rate is 6.1%.

71. Uneven development

1 The general wellbeing of individuals or societies.
2 People in developing countries often have very limited access to health care <u>as governments do not have the finances to support this. This results in a low number of doctors and hospitals per head of population. Consequences include high infant mortality rates and a short life expectancy. The opposite is true in developed countries.</u>
3 (a)

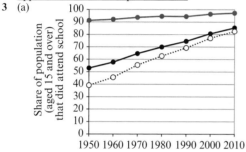

(b) In 2010, 76% of those over 15 attended school in developing countries compared to 94% in developed countries. Developing countries have less income to spend on education and money available is targeted at universal primary education, reducing the amount available for older students.
4 (i) Food security is essential for development. In less developed countries people are often inadequately fed and suffer from malnutrition, which affects their ability to work, therefore reducing their productively and life expectancy.
(ii) Most people in more developed countries have easy access to clean water, but many people in developing countries are forced to walk long distances to collect dirty, contaminated water. This leads to diseases such as cholera, which lowers life expectancy.

72. International strategies

1 This is where the government of one country voluntarily transfers resources, such as money and goods, to another country.
2 International aid can help to pay for imports, such as machinery and oil, which are vital to development. These will help to make agriculture more productive and to develop an industrial base, which will increase exports of more valuable secondary goods. International aid will also support the accumulation of enough capital to invest in industry and infrastructure, again increasing the economic base and allowing development.
3 (a) £8.63 bn + £8.80 bn + <u>£11.46 bn + £11.73 bn + £12.24 bn</u> = <u>£52.86 bn; £52.86 bn</u> ÷ 5 = <u>£10.57</u> bn
   (b) Aid from a single donor country (for example, UK) to a single recipient country (for example, Nigeria).
   (c) Africa contains many of the least developed countries so aid is essential for increasing the low agricultural production and increasing food security. Many regions, such as the Sahel, suffer from famine so aid is needed to reduce hunger and starvation.

73. Top-down vs bottom-up

1 Top-down development happens due to incentives and actions of governments and TNCs. The projects designed to increase development are usually large-scale ones, at national or regional level, such as HEP. Such projects increase the demand for labour during the construction phase and enable new skills to be learned, which help to extend a skilled labour force, and provide power for the future development of the country.
2 (a) Box 1: National governments play a relatively small role because <u>bottom-up development involves NGOs working with communities.</u> Box 2: Outside agencies such as WaterAid work with <u>villages and local communities.</u> Box 3: Bottom-up development schemes are planned and controlled by <u>local communities to help their local periphery area.</u> Box 4: Cheap compared to top-down development but <u>funded by the community.</u>
   (b) **D** Governments become over dependent on NGOs
3 Top-down projects can require large investments from TNCs that also provide knowledge and expertise for further projects to help countries' development. Large-scale projects, such as dams, can benefit thousands of people by providing fresh water and electricity. However, funding can come with conditions that might not benefit long-term development, such as the removal of trade barriers. Projects such as dams force people to move and lose their homes.

74. Location and context

1 *Example*

2 *Example* Emerging countries such as India are usually members of global groups including the World Trade Organization and the United Nations. <u>India is a federal democratic republic, which means that the President of India is head of the country but the Prime Minister of India is the head of the government. India is the world's largest democracy; the government is chosen as the result of a vote.</u>
3 **D** Climate
4 India's population is very socially divided into social ranks known as 'castes', which are assigned at birth. India's society therefore is based on a social conception, leading to inequality and hierarchy. This system is reinforced by traditional family life as the family is central to Indian society, and illegal castes such as the 'untouchables' are discriminated against.

75. Uneven development and change

1 The most economically advanced region or regions.
2 **D** Regions with lower levels of development
3 Box 1: The GPD of Goa is <u>high</u>. This means that Goa forms part of the core region because <u>it indicates that skilled workers are employed in industry and technology resulting in economic development.</u> Box 2: The GDP of Bihar is <u>very low</u>. This means that Bihar forms part of the periphery region because <u>low wages indicate an agricultural-based economy with low economic development.</u>
4 *Example* Increased mechanisation and industrialisation has meant that the contribution of agriculture to India's GDP has decreased from 58 per cent to 26 per cent. This change from a primary to a more broad-based economy has promoted development but has also encouraged a rapid rise in rural to urban movement. Poorly built and badly regulated shanty towns have been built around the edges of urban areas. These

have a negative impact on people's health and wellbeing due to the lack of basic services. Family values tend to break down in rural areas, as the younger generations – particularly men – move to towns, leaving the old and the young behind. Indian society is strongly based on family values, so this can be seen as a negative result of development.

However, the rapid growth of the quaternary sector, mainly due to the fastest growing telecom markets in the world, has meant that increased investment from TNCs has created over one million ICT jobs. This has not only helped development but has meant that India now has a highly trained work force with an international reputation.

The change from a primary to a more advanced economy has created a number of problems and helped to underline the gap between the core and peripheral regions of India. However, the economic benefits of the development of the secondary, tertiary and quaternary sectors will benefit India's economy, with greater stability in the short and long term.

76. Trade, aid and investment

1 International trade is exchanging goods, services <u>and capital, involving both importing and exporting, across international borders.</u>

2 (a) Imports exceed exports between March 2014 and March 2015: for example, there was a $12.2 bn deficit in July 2014. The imports and exports follow the same trend: when exports decrease, imports do as well and both fluctuate over time, although there is not a perfect match. The largest value of imports was in September 2014 ($43.2 bn) and the highest exports in March 2014 ($30.30 bn).

   (b) Development is linked to a rapid rise in imports and exports. As more manufactured goods and services are produced, the demand for imports rises as a result of consumer demand and the need for raw materials for manufacturing.

3 The amount of international aid given rises during a disaster such as a major flood or earthquake as international governments and charities such as Oxfam may provide major donations of food, money and services. Aid is also often linked to historic ties: developed countries will donate aid to former colonies, but they will decrease this aid as the country develops. Political change also causes variations in aid. A country that is regarded as corrupt will receive low levels of aid, but aid may increase if the level of corruption decreases.

77. Changing population

1 (a) **C** Population structure

   (b) In 1985, the population pyramid was triangular in shape; <u>by 2015 the side of the triangular shape had become steeper. In 1985, the base of the pyramid was broader, reaching 7 per cent for both male and female, but by 2015 the base had reduced to 6 per cent for both male and female. In 1985, the oldest age group shown on the pyramid was 75–79 years and this changed to 95–99 in 2015.</u>

2 *Example* In India, life expectancy has improved over the last 30 years, rising from 54 years in 1985 to 68 years in 2015. This is due to reduced infant mortality rate, as fewer children died before five years old because of improved diets and vaccinations against diseases such as measles and polio. There is also a reduced maternal mortality rate, with fewer mothers dying in childbirth as medical care has improved, especially in remote rural areas.

3 *Example* (a) Rapid urbanisation and increasing levels of education are causing an increase in the size of the middle classes, which is resulting in increased demands for consumer products and imports. This can result in a trade deficit.

   (b) Improved education will mean that the country changes from being economically based on primary products to having an economy based on secondary, tertiary and quaternary products. This will increase the value of exports, especially services, and help to make the country more developed.

78. Geopolitics and technology

1 Geopolitics is the impact of a country's human and physical geography <u>on its international politics and international relations.</u>

2 *Example* India has signed an agreement with Russia to supply the Indian army with missiles and fighter planes and to build more nuclear reactor plants. This will increase development as power supplies in India will expand, which will help with industrialisation and exports. However, India is in dispute with China about water resources on the Yarlung Tsangpo-Brahmaputra River. Dam construction could lead to a reduction in India's HEP production, limiting development.

3 (a) 3 points correctly added to diagram.
   (b) **D** Top 20%

4 *Example* In India, very remote and densely forested hill states such as Himachal Pradesh and Meghalaya are peripheral to development and relatively poor. Since 2004, the World Bank has been working with the government of India on India's National Road Program to expand rural road connectivity. This has helped to develop the economy as farmers can get produce to markets and improved technology can reach the villages. Children can now go to school, which will help development to continue in the future.

79. Impact of rapid development

1 *Example* Pakistan. Industrial development and increased transport, especially by road, result in the emissions of greenhouse gases such as CO_2, which reduce air quality and lead to climate change. Unregulated use of chemicals in agriculture and industry lead to water pollution, which reduces the purity of drinking water and can cause eutrophication, which reduces biodiversity.

2 Rise in consumerism strengthening the economy

3 Box 1: Rapid development means that people move to urban areas. This causes <u>unplanned shanty towns</u>. Box 2: These areas are often polluted and have poor air quality. This means that <u>respiratory diseases are common</u>. Box 3: There are no or very few services or roads for medical treatment access. Diseases such as <u>cholera result in short life expectancies.</u>

4 *Example* In India, rapid development is causing increases in population size and the demand for education. India has a literacy rate of 74 per cent (2011 census) and many of the illiterate are girls. Government incentives to reduce this means that the literacy rate for women between the ages of 15 and 24 has risen. Another factor is the number of children who are malnourished or undernourished in India. This is partly the result of the rapid growth of shanty towns where many children are raised in poverty, and partly due to uneven development within India. Peripheral hill states are likely to be isolated due to poor infrastructure. The government has introduced the 'midday meal scheme', mainly in state-run primary schools, to help reduce poor levels of nutrition. The Slum Rehabilitation Act was passed by the local government of the state of Maharashtra to protect people living in slum areas and to encourage the provision of services such as clean water in slum areas and shanty towns. In addition to these top-down schemes run by the government, there are bottom-up schemes, which are led by local people. Self-help schemes to improve shanty towns with properly built housing mean that people learn skills such as bricklaying, which improves living conditions and provides employment.

Resource management
80. The world's natural resources

1 Natural resources are materials which exist without the actions of humans: for example, fossil fuels.

2 Biotic resources are obtained from the biosphere and are capable of natural reproduction, such as plants and animals. However, abiotic resources are obtained from the lithosphere, atmosphere and hydrosphere. Examples include minerals and fresh water.

3 **A** Potentially inexhaustible; **D** Can be naturally replenished

4 (a) Total deforested 750 000 sq km / 100 = 7500 × 10 (Bolivia) = 75 000 sq km

   (b) Natural environments are exploited to obtain resources. For example, the rapid deforestation of tropical forests produces high-quality timber, which is a valuable export. Other reasons for exploitation are <u>for food – for example, fishing supplies important sources of protein – and clearance for agriculture. The demand for fossil fuels has led to extensive vegetation clearing for oil wells and overexploitation has resulted in a marked decrease in the reserves of fossil fuels available.</u>

81. Variety and distribution

1 (a) Southern (b) The Thames region contains much of Greater London and the Thames Valley, including large towns such as Reading. Therefore the population density <u>is very high, which means that water demand is also high, resulting in high volumes of water abstraction.</u>

2 Climate and latitude are both important factors. Higher precipitation and solar radiation levels near the Equator mean that tropical areas are very productive for extensive plantation agriculture. 30° N and S of the Equator, high solar radiation and very low precipitation result in little or no natural

vegetation and unproductive desert sandy soils. Agriculture in these regions is therefore restricted to nomadic herding, and crops can only be grown where irrigation is possible.

82. Global usage and consumption

1 (a) 520 Kcal
   (b) People in emerging and developing countries rely on their own produce for food or produce from their country. Therefore in a crisis there is a lack of food security and income to buy imported food, so the Kcal per person will be relatively low.
2 (a) Rapid industrial development in some countries
   (b) (i) Africa (ii) This continent includes a large number of countries that are either emerging or developing. Therefore <u>they do not have complex infrastructure, including electricity supplies, or high levels of technology, which means that energy use is low.</u>

Energy
83. Production and development

1 Non-renewable energy resources take a long time to form and therefore, once used, these are not replenished. Renewable resources are replenished over a relatively short period of time.
2 *Example* The development of coal extraction is expensive because of the high cost of developing mines and open cast pits. The actual extraction process is dangerous, particularly in deep mines, as gases such as carbon monoxide can kill miners. The process of mining can cause the release of methane, a greenhouse gas with a global warming potential 23 times higher than CO_2. Burning coal releases greenhouse gases such as CO_2 into the atmosphere.
3 There has been a steady increase in potential global wind power production between 2004 and 2014. <u>In 2004 the potential production was 48 gigawatts, but by 2014 this had risen by 322 gigawatts to 370. One of the largest increases was between 2013 and 2014, a rise of 51 gigawatts.</u>
4 Large reserves available
5 *Example* Wind energy is only produced when there is wind to move the blades, therefore it cannot be relied on for constant power generation.

84. UK and global energy mix

1 The range of energy resources of a country or region. The energy mix can be made up of either renewable or non-renewable resources or a mixture of the two.
2 In the 1970s, the UK had a large number of deep and open cast coal mines, which meant that coal was relatively cheap and readily available, so coal and oil provided over 90 per cent of energy consumption. However, coal became less economic to mine and concerns about greenhouse gas emissions have meant that renewables increased from 3.8 per cent of electricity in 2005 to 19.3 per cent in 2012. Government policy means that renewables will become more important in the future.
3 (a) USA
   (b) Countries with large resources of one type of energy do not have a true energy mix but depend on that resource as it is usually the cheapest. For example, New Zealand and Iceland rely on <u>geothermal energy. These countries import some petroleum for transport.</u>
4 Improvements in technology have greatly increased demand as households, especially in developed countries, own many electrical items like laptops, mobile phones and dishwashers.

85. Impacts of non-renewable energy resources

1 (a) Nuclear
   (b) (Coal) 42% + (oil) <u>1%</u> + (natural gas) 9% = 52%
   (c) All fossil fuels emit greenhouse gases when they are burnt. Burning coal produces a large amount of CO_2 and, although natural gas releases less CO_2, it is made up of methane. Greenhouse gases impact on the environment by contributing to global warming and climate change, resulting in raised global temperatures, melting ice sheets and glaciers, rising sea levels and the loss of biodiversity. Locally, ground and seawater can become contaminated by oil spillages, which result in severe pollution and the destruction of ecosystems.
2 Exposure to radiation
3 UK geologists believe there is enough gas in sedimentary rocks called shales to provide energy for the next 70 years. This shale gas can be recovered by fracking, which uses new technology based on injecting water and chemicals into tiny factures in the rock to release the gas, which will mean that the UK will not have to rely on imported gas.

86. Impacts of renewable energy resources

1 (a) Solar energy
   (b) **B** Reduces CO_2 emissions; <u>C Can generate enough energy for thousands of homes</u>
2 Disadvantages include disruption to both ecosystems and wildlife as large areas of vegetation are cleared.
3 Wind power can generate enough carbon-free energy to power thousands of homes.
4 (a) 4%
   (b) The increase is due to attempts to develop renewable energy to reduce emissions of CO_2 and other greenhouses gases, which are causing climate change. This development is seen as important as it helps to reduce global temperature increases and impacts, such as rising sea levels.

87. Meeting energy demands

1 Scientists suggest relying on fossil fuels could have irreversible impacts caused by climate change, such as sea level rises which <u>will displace many people in low-lying countries such as Bangladesh and destroy farmland, resulting in food shortages. Fossil fuels such as oil are non-renewable so these need to be carefully managed so that enough remains for the future.</u>
2 **B** Gas, oil and coal
3 *Example* All stakeholders, individuals, organisations and governments have different attitudes to the exploitation and consumption of energy resources depending on how much they are affected and the amount of pressure there is to manage resources.
   Individuals can be very active in promoting renewable resources, as in Germany where many people have solar panels on their roofs and put considerable pressure on local and national governments to develop renewables as opposed to nuclear, coal or oil-based fuels. However, some individuals feel that renewable energy is expensive compared to fossil fuels and do not support further development. Environmentalists concerned about the impacts of extensive wind farms on bird migration also object to development. Organisations may see both the financial and the PR advantages of managing their operations in a sustainable way. Companies such as McDonalds have a very positive attitude to transport as they convert used cooking oil into biodiesel for their delivery lorries. However, the increased costs of exploitation and consumption may be too expensive for smaller companies. The UK and 195 other nations pledged to reduce greenhouse gas emissions to zero at the 2015 Paris United Nations Climate Change Conference. Countries that see the immediate dangers of climate change and those with policies which include the sustainable use of energy are influencing other countries to invest in low carbon technology and local incentives to reduce petroleum consumption and exploitation, such as cycle hire.

88. China and Germany

1 *Example* The 2011 nuclear accidents in Japan resulted in Germany cancelling its nuclear reactor plans and investing in renewables instead, especially offshore wind farms. Germany plans to reduce greenhouse emissions by 40 per cent by 2022, to meet government and EU targets, and the development of renewable energy is important to achieving this.
2 Both the actual and predicted energy consumption for China increase much more rapidly than the consumption of the USA. <u>In 2010 both countries consumed 100 quadrillion British thermal units. However, by 2040, China's consumption is predicted to rise to 220 quadrillion British thermal units, while the USA's remains much lower at just over 100 quadrillion British thermal units.</u>
3 *Example* China is a rapidly developing country that contributes 29 per cent of global carbon emissions, the highest of any country. China passed a Renewable Energy Law in 2006, which aims to develop and extend the use of resources in a sustainable way. This was successful as in 2014 the Three Gorges Dam was generating 98.8 billion kWh of electricity, approximately the same as formerly produced by 49 million tons of coal. In addition, solar power is being developed in the Gobi Desert to produce enough energy for 1 million homes, and the Chinese government introduced laws in 2015 to reduce the use of coal in urban areas where smog is a major health hazard.
   Germany, a developed country, produces more than 28 per cent of its energy from renewables and aims to reduce greenhouse gas emissions by 40 per cent by 2022. Germany has invested heavily in offshore wind farms and 'solar parks',

such as Bavaria Solarpark, which is designed to produce 215 million kWh of power over the next 20 years, and reduce CO_2 emissions by 100 000 tons over the next 30 years. Germany has therefore been highly successful in the sustainable management of energy.

China has been very successful recently in developing and managing sustainable energy resources. For example, China reduced its coal consumption by 6 per cent and total emissions by 5 per cent in the first half of 2015. Germany has also been highly successful and is the world's leader in reducing carbon emissions, aiming to cut these by 80 per cent by 2050.

Water
89. Global distribution of water
1 Water without salt.
2 (a) Glaciers and permanent snow cover.
   (b) Most of the fresh water is in stores that makes it impossible to use. For example, nearly 70 per cent <u>is stored in glaciers and permanent snow cover and therefore unavailable. Some of the 30.8 per cent stored as groundwater is permafrost, so is also unavailable, leaving only 0.3 per cent in rivers and lakes, which is more readily available.</u>
3 Water surplus occurs when precipitation exceeds evaporation and transpiration.
4 *Example* The Sahel. The Sahel has about 200–600 mm of rainfall a year, which occurs between May and September due to monsoon season. However, the monsoon is unpredictable and therefore sometimes the amount is lower. As mean monthly temperatures vary between a maximum of 36 °C and a minimum of 18 °C, water loss due to evaporation and transpiration is significantly higher than rainfall, causing a water deficit.

90. Changing water use
1 **A** Increased
2 Asia includes developing countries such as India, where rapidly increasing populations are increasing food demands. More water is needed for <u>irrigation as the amount of irrigated farmland has increased from approximately 22.6 million hectares to 90 million hectares.</u>
3 (a)

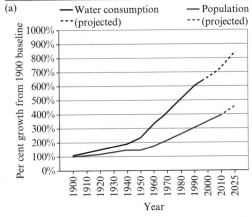

   (b) Global population has increased from 200 per cent in 1970 (based on 1900 being 100 per cent) to just over 300 per cent by 2010. In the same period, water consumption increased from 500 per cent to 700 per cent. Global population increase is an important factor behind increased demand for water, but other factors, such as developments in technology, mean that water-using appliances such as dishwashers are more commonly used in developed countries, increasing water demand.

91. Water consumption differences
1 Mean water consumption is the total of all the water used in the country divided by its population.
2 Box 1: Developed countries, e.g. the USA, use drip-feed sprinklers which reduce water use but <u>automated watering systems use a lot of water, increasing total usage.</u> Box 2: Emerging countries, e.g. India, rely on domestic agriculture for food production which means that <u>the amount of water used for irrigation is high.</u> Box 3: Developing countries, e.g. Angola, tend to use basic irrigation systems such as hand pumps so their water use is <u>relatively low.</u>
3 **A** People bathe in streams and rivers; **D** People's homes lack piped water

4 Developed countries can use a large volume of water for industry; in some cases this can be 80 per cent of the water available. The largest single use of water by industry is for cooling in thermal power stations, so this may decline as more countries use renewable energy. Industry is not such a significant user in many developing and emerging countries because many industries are small scale, such as basket weaving, and do not require large volumes of water.

92. Water supply problems: UK
1 Higher rainfall tends to occur in areas of lower population density, such as the Highlands of Scotland and north Wales, so these areas do not have a water deficit or a supply problem. However, <u>lower rainfall occurs in the south and east of the UK, which contain areas of very high population density, such as London.</u>
2 (a)

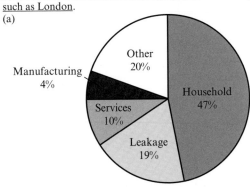

   (b) There is a high loss due to leakage as much of the ageing infrastructure means old water pipes are unable to cope with the higher water pressure needed to supply large urban areas.
3 (a) Seasonal imbalances occur because the UK receives much of its rain during the winter and the lowest amounts of rainfall during the summer months, creating a seasonal imbalance. However, some regions, such as East Anglia, have a small variation between summer and winter rainfall, so the seasonal imbalance here results from demand, which is much higher during the summer due to irrigation for farming.
   (b) Low rainfall in the summer can cause low reservoir levels and restrictions on water usage.

93. Water supply problems: emerging or developing countries
1 (a) Untreated water has not been put through a process to remove contaminants or organisms, such as bacteria.
   (b) Without safe water it is very difficult for people to lead healthy and productive lives. If people are not productive, food production is low and this means that <u>their ability to farm or carry out other work is limited, leading to a low calorie intake and an unbalanced diet. This makes them more susceptible to diseases and more likely to die from the contaminates and parasites found in untreated water.</u>
   (c) **C** Africa
2 (a) The Sahel region of Africa.
   (b) Low rainfall in these regions can result in drought. People do not have enough water to irrigate their crops, which results in famine and malnutrition.

94. Attitudes and technology
1 (a) Desalinisation.
   (b) One advantage of this method is that it reduces the demand for groundwater extraction, which means that there is less impact on ecosystems <u>and wildlife as rivers are less likely to become dry. One disadvantage is that desalinisation plants are expensive to run as they need frequent cleaning, because fish and waste materials may be sucked in as water is collected for treatment.</u>
2 (a) Map completed using correct colours from key.
   (b) Regions of predicted severe water shortage form a band stretching west to east between the Equator and the Tropic of Cancer in the Mediterranean and Caspian Sea areas. The band stretches from Spain and Portugal in the west to Kazakhstan in the east.
3 *Example* The Las Vegas local government is very active in encouraging sustainable water management through a Water Resource Plan, which encourages local residents to reduce their water consumption by incentives (for example, paying them to replace lawns with desert gardens). Although some

residents are happy to do this, there is opposition from those who wish to maintain their grass. This varied response is also true of organisations, with some casinos operating water recycling schemes, while other organisations have not adopted such policies.

95. Managing water

1 C Water abstraction is higher than the amount available
2 Allowing controlled development in a way that meets present needs without affecting the ability of future generations to meet their own needs.
3 Water is essential for life, and as global populations increase water resources are increasingly under stress. Therefore <u>they need to be managed in a way that conserves supplies for future generations. If water resources are used at the current unsustainable rate, it is estimated that by 2025 two-thirds of the world's population will live in water-stressed countries and that by 2030 global demand will be 40 per cent greater than supply.</u>
4 (a) 12% (accept 11–13%)
   (b) *Example* The local government accepts the need to maintain high water levels in Lake Mead as the water levels are currently falling below the level of the pipes that take water to Las Vegas. They are therefore encouraging sustainable water use, such as replacing heavily water-thirsty grass with desert gardens that do not need watering. However, other users are reluctant to give up their lawns and swimming pools, and do not agree with the management of what they regard as their right. Environmental groups support sustainable management as alternative plans include abstracting large volumes of groundwater, which would have a negative impact on dependant ecosystems.

96. UK and China

1 (a)

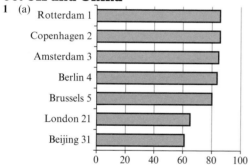

Water management index score

   (b) Cities in developed countries have more finance available because <u>of higher GDP and income levels, which allows greater investment in sustainable water projects. Water resources are more reliable than in developing or emerging countries, which means it is easier to enforce legislation that encourages water recycling.</u>
2 *Example* In the UK, water companies are encouraging sustainability by treating wastewater and recycling it for domestic and industrial reuse, reducing the amount of water extracted from the ground or taken from surface supplies. Water companies are also encouraging installation of water meters, which make users more aware of the amount of water they are using and help to reduce water usage.
3 *Example* China has an imbalanced water supply, with the greatest demand for water in the east of the country where supplies are lowest. Since 2011, China has developed methods to manage water use in urban and agricultural areas more sustainably, although some schemes have been more successful than others. Although there is sustainable management in Beijing, which recycles 85 per cent of its wastewater, and new recycling plants are being developed there, the majority of cities do not have metering equipment to monitor abstractions from rivers, therefore unsustainable levels of abstraction are frequent. In rural areas, despite incentives such as planting drought-resistant crops and controlled irrigation, there is no water abstraction metering, so river and groundwater supplies continue to be overused. Although China is making progress in managing water resources in a sustainable manner, this is at present limited to large urban areas and in particular Beijing, with very limited national management.

Extended writing questions
97. Paper 2

A02

Developed countries (e.g. the UK)
*   There is a national water imbalance, most rainfall occurs in the northern and western regions where population density is low. Areas such as south-east England with a very high population density (407 people per sq km) have low rainfall totals.
*   There are seasonal imbalances in rainfall and water demand – rainfall is low in the summer in the east and south-east of the UK when demand is high for agriculture and gardening – this is also the area of highest population density.
*   Much of the infrastructure is over a hundred years old which leads to leaking pipes – it is estimated that there is a national leakage rate of 24%.

Developing or emerging countries
(e.g. China/Chad/Brazil/India)
*   The lack of water treatment plants and rapid urbanisation means that many people do not have access to safe drinking water. In the favelas in Brazil, 92% of households have access to safe water but in rural areas this is more of a problem as only 65% of homes have a water supply.
*   Water courses are polluted as these act as sewage systems and rubbish dumps; this is especially true in many developing and emerging cities. In Dharavi, a very poor area of Mumbai, India, some districts depend on wells which provide polluted water.
*   Some areas such as the Sahel region of Africa have a low and unpredictable rainfall pattern; this is becoming even more unpredictable as monsoon patterns change as a result of global warming.

A03
*   Both developed and developing/emerging countries have water problems; however, developing countries have the finance and the technology available to alleviate the problems.
*   However, even in developed countries such as the UK lack of investment means that water leakage is a major problem.
*   In developing or emerging countries, water supply problems are only being addressed in major urban areas and not as a whole country incentive.
*   Global climate change will change rainfall patterns and create new problems for countries at all levels of development.

COMPONENT 3: GEOGRAPHICAL INVESTIGATIONS
Fieldwork: coasts
98. Formulating enquiry questions

1 Box 2: Use a range of fieldwork techniques and methods; Box 4: Analyse and explain data; Box 6: Evaluate data and data collection methods.
2 (a) How and why does the beach profile change along a stretch of coastline?
   (b) The OS map shows that Swanage Bay is accessible. There are no high cliffs in the southern section and a road runs along the beach. Therefore it is a stretch of coastline that can be easily investigated using a number of sites.
3 (a) Longshore drift
   (b) The map indicates that this is a drift-aligned beach as the groynes are perpendicular to the beach. This means that beach material is deposited where waves break at an angle to the coast. The swash therefore occurs at an angle, but the backwash is perpendicular to the coastline. Beach sediment is therefore transported along the beach by longshore drift and will build a beach profile that can be measured as part of the investigation.
4 Groynes

99. Methods and secondary data

1 (a) C Data collected by someone else
   (b) The geological map helps to explain the headland and bay formation of Swanage Bay. The headlands are formed of more resistant chalk and limestone which eroded more slowly, while the bay is formed of softer clays and sands which eroded more rapidly.
2 Random sampling is the least biased of the sampling techniques because it is not subjective so each site <u>has an equal chance of being selected, meaning that the results will be unbiased. However, all the sites may be clustered together, so an investigation that is based on changes along a stretch of coastline would not produce data that would prove or disprove a hypothesis.</u>

3 Qualitative fieldwork involves collecting information on thoughts and opinions, usually by carrying out interviews or questionnaire surveys. However, quantitative fieldwork involves collecting numerical data or data that can be used statistically, such as pedestrian or traffic counts.

100. Working with data
1 (a) Photographs give a clear visual impression of the general area and sites used to collect data, and <u>are an excellent way of showing features such as the build-up of sediment on either side of a groyne, which is difficult to explain verbally.</u>
  (b) Pie charts may have a large number of small segments, such as less than 5 per cent, which can make the chart difficult to read and interpret.
2 (a) There is an overall increase in height in the beach with distance from the sea, increasing from 0 to 4 metres high. However, this is not a smooth curve: at 20 metres from the sea, the profile levels off and at 54 metres from the sea, the profile becomes steeper.
  (b) The increase in height is due to sediment deposition over time by longshore drift. The steeper section of the profile at 54 metres from the sea is a storm beach, caused by high-energy waves depositing larger sediment.

Fieldwork: rivers
101. Formulating enquiry questions
1 Box 2: Use a range of fieldwork techniques and methods; Box 4: Analyse and explain data; Box 6: Evaluate data and data collection methods.
2 (a) How and why do river channel characteristics vary along a stretch of the River Barle?
  (b) The OS map shows that the River Barle is accessible. There are no steep river cliffs indicated and a road or footpath runs along much of the river. Therefore it is a stretch of river that can be easily investigated using a number of sites.
3 (a) The Bradshaw model.
  (b) The Bradshaw model shows theoretical changes of a number of variables from the source to the river mouth, but the theories can also be applied to a stretch of a river. For example, the model indicates that the channel depth decreases with distance downstream but the channel width increases. These changes result from changes in the velocity and gradient of the river and can be measured as part of the investigation.
4 Bridge at Dulverton

102. Methods and secondary data
1 (a) C Data collected by someone else (b) The flood risk map shows that the river exceeds bank full conditions and indicates the extent of the river floodplain. It also shows that the river channel increases in width and depth when velocity and discharge are high and can be used to explain river channel features such as river cliffs.
2 Random sampling is the least biased of the sampling techniques because it is not subjective so each site <u>has an equal chance of being selected, meaning that the results will be unbiased. However, all the sites may be clustered together, so an investigation that is based on changes along a stretch of river would not produce data that would prove or disprove a hypothesis.</u>
3 Qualitative fieldwork involves collecting information on thoughts and opinions, usually by carrying out interviews or questionnaire surveys. However, quantitative fieldwork involves collecting numerical data or data that can be used statistically, such as pedestrian or traffic counts.

103. Working with data
1 (a) Photographs give a clear visual impression of the general area and sites used to collect data, and <u>are an excellent way of showing features such as the height of the river relative to bank full conditions, which is difficult to explain verbally.</u>
  (b) Pie charts may have a large number of small segments, such as less than 5 per cent, which can make the chart difficult to read and interpret.
2 (a) The river channel is asymmetrical. The river channel bed dips in an easterly direction at an angle of approximately 45°. At a metre away from the west bank, the deepest section of the channel – the eastern bank – rises steeply at an angle of approximately 75°.
  (b) The shape is due to variations in velocity. Faster velocity causes more energy on the eastern side of the river

channel, resulting in erosion, while deposition occurs towards the west bank, where river velocity is lower.

Fieldwork: urban
104. Formulating enquiry questions
1 (a) Analysing and explaining data.
  (b) <u>The evaluation stage should include a review of the data collection methods,</u> data presentation methods and any conclusions made. This is because evaluating each stage allows you to establish the validity of the overall investigation.
2 (a) Questionnaire survey.
  (b) Important observations that are not asked must be included, such as the estimated age of the person answering the questions. The questions should be asked in a logical order. Closed questions are quicker and easier to ask and record than open questions, but may not give as much information, so the overall purpose of the survey needs careful consideration.
  (c) The street appears to be very similar along its length. It is formed of the same design of terraced housing with cars parked on either side of the road, giving a relatively narrow space in the middle. There does not appear to be a variation in environmental quality, so an environmental survey would record the same results at each site.

105. Methods and secondary data
1 (a) *Example* Questionnaire survey.
  (b) *Example* One advantage of using stratified sampling when carrying out a questionnaire survey is that each age group within the population is properly represented, so this gives an accurate <u>representation of the views of the community or group. However, the main disadvantage is that it is unlikely that the correct number of responses for each age group can be obtained, giving biased results.</u>
2 (a) **D** Secondary
  (b) *Example* The students could use the information to decide how many people of each age to ask when carrying out the questionnaire survey, therefore getting responses from a representative sample of the population of Fenham. For example, nearly 50 per cent are aged between 25 and 64, so this age group should be targeted to answer 50 per cent of the questionnaires.
3 The students carried out their data collection at a time when the largest age group (25–64) would have been at work. Therefore they would have been unable to question the section of the population that was most significant, in terms of numbers, so the results would be extremely biased and unrepresentative. In addition, carrying out the survey in September meant the 15–24 age group would be at school or work, biasing the results towards the over 65 age group, which only make up 14.6 per cent of the population.

106. Working with data
1 (a) Scattergraphs can only be used show relationships between two variables, so are inappropriate in this case as there are three sets of data.
  (b) A histogram.
  (c) Data is grouped into categories such as age groups, so it is impossible to read exact values, such as the answer given by someone aged 16.
2 (i) Are there enough green spaces in Fenham ward? (ii) *Example* <u>Are the pavements well maintained?</u>
3 *Example* I tested the methods I used to collect data by carrying out a pilot study in the area around the school. I was able to try out the questions in the questionnaire survey about the quality of the environment and I changed these as a result, making my actual data collection more focused and accurate. I also used census data to decide how many people of each age group to ask (stratified sampling). Therefore my data collection was representative of the population of the area I visited. I also pilot-tested my recording sheet for land use function and decided to use systematic line sampling along the transect as this removed most of the bias.
I carried out the actual data collection on a Saturday in order to get a range of age groups for the questionnaire. This was very successful as I obtained a representative sample, based on my secondary data. However, the sampling method used for land use was less successful as line systematic sampling meant that some land use changes were not recorded, and other sampling points could not be used because they were in dangerous places, such as the middle of roads. My results were therefore very biased.

Fieldwork: rural
107. Formulating enquiry questions
1 (a) Analysing and explaining data.
  (b) <u>The evaluation stage should include a review of the data collection methods,</u> data presentation methods and any conclusions made. This is because evaluating each stage allows you to establish the validity of the overall investigation.
2 (a) Traffic survey.
  (b) The form should include space to note the time, date and day of week of the survey as these factors can help explain the traffic movements recorded. There should be clear columns for each direction of flow to avoid these becoming confused. There should be separate columns for each type of traffic, such as cars, buses and lorries, with a column headed 'Others' for unexpected traffic types.
  (c) The location should be rejected for safety reasons. There are no pavements, so anyone collecting traffic survey data would have to stand in the narrow road next to a wall, which would be extremely dangerous.

108. Methods and secondary data
1 (a) *Example* Questionnaire survey.
  (b) *Example* One advantage of using stratified sampling when carrying out a questionnaire survey is that each age group within the population is properly represented, so this gives an accurate <u>representation of the views of the community or group. However, the main disadvantage is that it is unlikely that the correct number of responses for each age group can be obtained, giving biased results.</u>
2 (a) **D** Secondary
  (b) *Example* The students can use the information to decide how many people in each age to ask when carrying out the questionnaire survey, therefore getting responses from a representative sample of the population of Llanbedr. For example, nearly 50 per cent are aged between 25 and 64, so this age group should be targeted to answer 50 per cent of the questionnaires.
3 The students carried out their data collection at a time when the largest age group (25–64) would have been at work. Therefore they would have been unable to question the section of the population that was most significant, in terms of numbers, so the results would be extremely biased and unrepresentative. In addition, carrying out the survey in September meant the 15–24 age group would be at school or work, biasing the results towards the over 65 age group, which only make up 24.3 per cent of the population.

109. Working with data
1 (a) Scattergraphs can only be used to show relationships between two variables, so are inappropriate in this case as there are three sets of data.
  (b) A histogram.
  (c) Data is grouped into categories such as age groups, so it is impossible to reads exact values, such as the answer given by someone aged 16.
2 (i) Are there enough green spaces in Llanbedr? (ii) *Example* <u>Is footpath erosion a problem in the area?</u>
3 *Example* I tested the methods I used to collect data by carrying out a pilot study in the area around the school. I was able to try out the questions in the questionnaire survey about the quality of the environment and I changed these as a result, making my actual data collection more focused and accurate. I also used census data to decide how many people of each age group to ask (stratified sampling). Therefore my data collection was representative of the population of the area I visited. I also pilot-tested my recording sheet for traffic flow and decided to carry out the survey every half hour for 5 minutes.
  I carried out the actual data collection on a Saturday in order to get a range of age groups for the questionnaire. This was very successful as I obtained a representative sample, based on my secondary data. However, the traffic survey was less successful. The road through the village was very narrow and it was difficult to find a safe place to carry out the survey. Cars parked along the road and the narrow bridge meant that there was traffic congestion, so only one or two vehicles passed me during the recording time. My results were therefore very biased.

UK challenges
110. Consumption and environmental challenges
1 (a) If a country, region or city is overpopulated it has too many people for its resources, such as the amount of food, materials, and space available.
  (b) The UK's population is predicted to increase steadily from 65 million in 2015 to more than 77 million in 2050. This is an increase of <u>12 million people in 35 years and the demand for and consumption of resources – especially food and energy – are likely to exceed the supply available in the UK. This will lead to increased imports of food and the development of renewable resources, such as wind and tidal energy in the future.</u>
  (c) **A** Reduction in ground water supplies; **C** Loss of habitats
2 (a) Car sharing.
  (b) Sustainable transport systems help to reduce greenhouse gases, such as CO_2, therefore reducing the amount of pollutants that cause climate change and helping to control the probable negative impacts on UK's weather and climate. Sustainable transport systems such as cycling do not use fossil fuels, which helps to reduce the use of petroleum and reduces the amount of oil that the UK imports.

111. Population and economic challenges
1 (a) Varying regional rates of economic growth
  (b) One proposed solution is improving transportation links with the north to encourage business development. The proposed HS2 rail link to Manchester and Leeds will create a large number of jobs, 70 per cent of which will be outside London. The creation of a 'Northern Powerhouse' will attract investment into northern cities and towns by bringing together the major northern economies, such as Manchester.
2 **A** Loss of agricultural land; **D** Affects ecosystems
3 There appear to be no reliable methods of monitoring migration. Despite the problems in obtaining a visa to come to the UK, <u>there does not seem to be a method to check if someone is still in the UK when it expires. There are also an unknown number of illegal migrants who arrive each year in lorries or through small, unregulated ports. The statistics are therefore probably unreliable.</u>

112. Landscape challenges
1 (a) **A** Dartmoor; **C** Cairngorms
  (b) National Park authorities use education to help visitors understand the need for conservation. For example, information boards allow people to enjoy National Parks while appreciating the need for management. Sustainable transport schemes, such as bike hire and shuttle buses, also help to reduce air and noise pollution, therefore helping to conserve fragile ecosystems within the parks.
2 (a) There has been a marked increase in the number of closures. Between 1983 and 1989 there was only one closure a year or less, but after 1990 this increased to <u>approximately three a year, rising dramatically in 2001 to 25 a year. Overall, there are more closures due to tidal flooding than fluvial flooding.</u>
  (b) One approach involves hard engineering. This relies on erecting barriers, such as sea walls, which prevent floodwater from reaching land areas. Soft engineering and managed retreat allow more natural methods of controlling flooding to occur. Managed retreat allows areas of low-value land to be flooded, therefore reducing protection costs and creating valuable habitats, such as salt marshes.

113. Climate change challenges
1 (a) **B** + 4 °C in southern England.
  (b) Computer models suggest that climate change resulting from global warming will increase the number of wet winters, and that there will be a higher frequency of intense downpours.
2 (a) (i) People can use public transport to reduce their carbon footprint (ii) <u>People can install loft insulation to reduce heat loss from homes, which reduces fossil fuels usage.</u>
  (b) The UK government is promoting the use of more sustainable practices, such as hire bikes in urban areas and national parks, as this can help to reduce emissions of greenhouse gases associated with climate change. The government could also increase investment in renewable energy, such as wind farms, to reduce dependence on fossil fuels, therefore reducing greenhouse gas emissions.

Extended writing questions
114. Paper 3 (i)

1 The student only investigated three sites. Therefore <u>it is unlikely that they collected representative data for a river 6.5 km in length.</u> Ten sites would have offered a more valid comparison to Bradshaw's model and would have allowed statistical tests to be carried out. The locations of the data collection sites are not given. These may have been close together at the source or mouth. Clustering would not give a representative picture of downstream changes. The student should have used a sampling method such as stratified sampling to select sites and reduce bias.
<u>The student does not state if the width and depth are mean figures or single measurements, but there is not enough data</u> to make conclusions about the entire river changes. The very basic conclusions drawn by the student lack evidence and justification.
<u>Overall even the basic conclusions drawn by the student cannot be justified.</u>

2 The student only investigated three sites. Therefore <u>it is unlikely that they collected representative data for an extensive coastal location.</u> Ten sites would have offered a more valid comparison when investigating coastal processes such as longshore drift, and would have allowed statistical tests to be carried out. The locations of the data collection sites are not given. These appear to have been close together near to one groyne. Clustering would not give a representative picture of variations in beach profile. The student should have used a sampling method such as stratified sampling to select sites and reduce bias.
<u>The student does not state if sediment size variations are mean figures or single measurements, or if sampling was used, but there is not enough data to make conclusions about</u> changes along a beach profile. There is some reference to geology, but this is not developed to help explain the type and size of beach sediment present. The very basic conclusions drawn by the student lack evidence and justification.
<u>Overall even the basic conclusions drawn by the student cannot be justified.</u>

115. Paper 3 (ii)
Example

Climate change can occur naturally, for example, due to variations in the amount of solar energy reaching Earth. However, most of the present rapid changes in global climate are due to an increase in the greenhouse effect, which is when the amount of heat retained by Earth's atmosphere increases due to human causes. Most of the information about the impact of climate change emphasises the negative effects, such as sea level rise and global water shortages.

Models of change for the UK also tend to indicate negative impacts. The land near or below sea level in East Anglia is predicted to be flooded, causing the loss of valuable farmland, which will reduce food security in the UK and also cause the loss of settlements. This will cause social problems as communities are lost, and will increase the pressures in other parts of the UK that are already overpopulated, such as London and the south-east. However, there will be benefits. An increase in temperatures – estimated at 2 °C to 5 °C by 2080 – will encourage more tourism in the UK, which will benefit the economy and local employment, especially in coastal areas. The increased income might be used to extend coastal defenses to protect important ecosystems and urban areas. Climate change will also allow agricultural crops such as tea and wine to be grown more widely, reducing the need for imports.

Figure 1 suggests some additional benefits. Warmer sea temperatures might mean that coldwater fish such as cod move further north, but other fish species will take their place, increasing fishing opportunities. Drier summers indicate drought, especially in the east of the UK, but models suggest that the UK will have wetter winters. This water could be stored for use during the summer, reducing the drought risk and the shortage of drinking water.

Although Figure 1 emphasises the negative impacts, there are a number of possible methods to reduce the severity of climate change on the people and landscape of the UK.

GEOGRAPHICAL, MATHEMATICS AND STATISTICS SKILLS
116. Atlas and map skills

1 (a) **A** Dispersed; **D** Linear
  (b) Deciduous forests and woodlands are mostly located north of the Tropic of Cancer. They form a dispersed linear band <u>stretching from the east coast of North</u> America, through Northern Europe to Japan in the east. There are isolated areas of deciduous woodland in the southern hemisphere, for example the east coast of Australia.

2 (a) (i) Country boundaries (ii) Climate zones
  (b) The height and shape of the land's surface.

117. Types of map and scale

1 (a) **A** The size of an area on a map compared to real life.
  (b) (i) <u>1:250 000</u> (ii) 1:50 000
2 **C** 6 km
3 (a) Isolines are lines that join together points of equal value.
  (b) OS maps (contours).

118. Using and interpreting images

1 (a) **B** Oblique aerial
  (b) It shows more of the area than a ground-level photograph.
2 (a) The detail in the background may not <u>be very clear.</u>
  (b) Satellite images are images of all or part of the earth taken using artificial satellites. These images can be used for mapmaking and weather forecasting.
3 Abrasion.

119. Sketch maps and annotations

1 Sketch maps can be annotated to add detailed information and explanations, such as changes of land use in a settlement.

2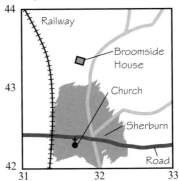

3 1 = E, 2 = C, 3 = A, 4 = D, 5 = B

120. Physical and human patterns

1 Annotated field sketches can be used because patterns such as the distribution of settlements can be clearly shown and labeled, so that the distance and spacing between the settlements can be analysed easily.
2 (a) Alnwick has a nucleated settlement shape. It is clustered around main roads and sandwiched between the A1 and the B6341 to produce an almost square shape.
  (b) It would be difficult to expand Alnwick because of the <u>A1 road to the east and the south east</u> of the settlement. <u>This road prevents any development in these directions. There is farmland to the east of the A1, therefore land is valuable for primary industry. The River Aln is to the N and NE, which prevents development; high land to the W and SW would be difficult to build on. Woodland to the north west of Alnwick reduces urban growth point this direction.</u>

121. Land use and settlement shapes

1 The key and symbols used on an OS map may show features linked with urban or rural land, such as railways and major roads. Different types of land use, such as farmland or woodland, may be shaded differently. Contours may also indicate flatter land, more suitable for building and farming.
2 (a) **C** Spread out
  (b) (i) 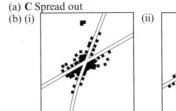 (ii)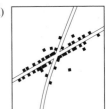

3 (a) Sidford is a nucleated settlement clustered around roads. Not all of the settlement can be seen on the map extract but it appears to have a rectangular shape.
  (b) The land use is very rural. Brook Farm is located to the <u>NW of Sidford in 1290 and is typical of the scattered farms indicating an agricultural area. A rural landscape</u>

is also suggested by a lack of built-up areas and transport routes are limited to roads less than 4 m wide, except the A375, which suggests limited use by a rural population. The only settlement in the area is Sidford.

122. Human activity and OS maps

1 (a) Museum

(b) Tourist information centre in GR 1813

2 (a) **C** Coniferous woodland

(b) Bus station

3 Shilbottle is surrounded by a rural landscape. There are many farms around the area, which show it is a rural agricultural area. For example, South East Farm located to the south east of Shilbottle. The village is relatively small and the houses and roads are not tightly packed together, indicating a rural area. There are limited services: only a church and a public house are shown on the map, which indicates a rural area.

4

Large factory buildings

Non-residential area

123. Map symbols and direction

1

•144   CH

2 (a) South-west (or North-east)

(b) Mixed deciduous and coniferous woodland.

(c) M50

124. Grid references and distances

1 (a) Deciduous

(b) 113911

(c) 4 km

(d) **B** Claypitts Farm

2 **C** 091923

3 (a) 6 km

(b) Telephone

(c) 112909

(d) 087919

125. Cross-sections and relief

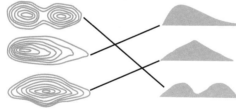

2 (a) **D** 99 m

(b) 70 m

3 (a)

A   Top of hill   B

Height of land in metres: 840, 820, 800, 780, 760

(b) The western side of the hill rises evenly in increments of 20 m. The top of the hill is relatively flat and is approximately 840 m above sea level. The downward slope off the top of the hill is a mirror image of the left (west) side. It then flattens out between 800 m and 780 m and then steeply declines to 760 m.

126. Graphical skills 1

1 (a)

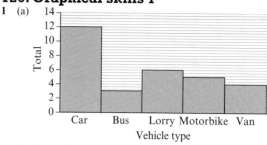

Total vs Vehicle type (Car, Bus, Lorry, Motorbike, Van)

(b) The traffic data cannot be shown as a line graph as it does not show a change over time. Line graphs are used to show trends or patterns to see if there is a correlation between two sets of data. There is only one set of data (vehicle numbers) in this case.

2 The graph shows a strong positive correlation. The larger the population of a settlement, the greater the number of services it has. For example, Hambury has a population of 125 people and three services and Wooferlow has a population of 1500 and 19 services. Shelton is a slight anomaly with a population of 1150 and 8 services.

127. Graphical skills 2

1 It is difficult to draw the symbols to the correct scale, making these maps hard to construct accurately.

2 (a) A wind rose diagram shows wind direction and speed vary over a period of time.

(b) Winds were recorded for all directions but the most frequent wind direction was south, about 18% of total winds. Other frequent wind directions were SE, SSE and N. There were very few NW winds.

128. Graphical skills 3

1 (a) (i) Data is shown for males and females (ii) Data is shown according to age.

(b) Developing country

(c) Developing countries populations are frequently pyramid-shaped; they have a large proportion of children, as shown by the pyramid for Pakistan which is bottom heavy.

2 An advantage of choropleth maps is that they show spatial variations over a geographical area. A disadvantage is that they suggest sudden changes between areas but the variations may be gradual.

3 Flow diagrams are easy to construct. As they show the direction of movement they give an excellent visual impression.

129. Numerical and statistical skills 1

1 70%

2 (a) 24.5%

(b) 30%

3 51.5

130. Numerical and statistical skills 2

1 (a) The median is the middle value of a set of numbers.

(b) 13.18 cm

(c) $(n + 1) \div 4$

(d) 'lower quartile' added to table, in row 7, second column

2 (a) $19 - 7 = 12$; 12 is the quartile range

(b) The IQR may be a better measure of spread than the range, as it is not affected by anomalies. It shows the range of pebble size on the beach, ignoring the very large and very small sizes which might distort the result.

3 **A** The number in the middle of a group

Published by Pearson Education Limited, 80 Strand, London, WC2R 0RL.

www.pearsonschoolsandfecolleges.co.uk

Copies of official specifications for all Pearson qualifications may be found on the website: qualifications.pearson.com

Text © Pearson Education Limited 2017
Typeset, illustrated and produced by TechSet Ltd.
Picture research by Caitlin Swain
Cover illustration by Miriam Sturdee

The right of Alison Barraclough to be identified as author of this work has been asserted by her in accordance with the Copyright, Designs and Patents Act 1988.

First published 2017

20

10 9 8 7 6 5

British Library Cataloguing in Publication Data
A catalogue record for this book is available from the British Library

ISBN 978 1 292 13373 7

Acknowledgements
Content written by Rob Bircher, Michael Chiles and Anne-Marie Grant is included.

The author and publisher would like to thank the following individuals and organisations for permission to reproduce copyright material:

Photographs
(Key: b-bottom; c-centre; l-left; r-right; t-top)

Alamy Images: Chris Craggs 26, D Core / Ocean 5, David Lyons 8, epa european press agency b.v. 79, EYESITE 18, Ian Dagnall 21, Ian Woolcock 6, John James 104, Jon Sparks 3t, Jozef Mikietyn 107, Midland Aerial Pictures 4, 17, Mike Greenside 118t, Nick Hawkes 100, Paul Broadbent 12, Pearl Bucknall 29, PURPLE MARBLES YORKSHIRE 103, robertharding 19, Stanislav Halcin 30, Travel Scotland - Paul White 14, www.mjt.photography 11; Fotolia.com: lenisecalleja 118b, stock_alexfamous 95; Roger Davies: 3b; Shutterstock.com: Dai Mar Tamrack 28, NCG 2, PaulPaladin 51

All other images © Pearson Education

Figures
Figure on page 39 adapted from Sahel precipitation anomalies 1950-2013 Sahel Precipitation Index (20-10N, 20W-10E), 1900 – May 2015 (http://research.jisao.washington.edu/data_sets/sahel/), Todd Mitchell November 2013 Joint Institute for the Study of the Atmosphere and Ocean NOAA | NOAA Cooperative Institutes DOI 10.6069/H5MW2F2Q, Joint Institute for the Study of the Atmosphere and Ocean at the University of Washington; Figure on page 61 from Areas of deprivation in Birmingham Crown Copyright © Crown Copyright Indices of deprivation 2015 – Gov.uk 2015 https://www.birmingham.gov.uk/deprivation-areas Department for Communities and Local Government, Department for Communities and Local Government © Crown Copyright. Contains public sector information licensed under the Open Government Licence v3.0.; Figure on page 62 from Retail locations in Birmingham, http://retailbirmingham. co.uk/new-birmingham-shopping-map-2015/ Retail BID; Figure on page 77 from Population pyramids for India, 1985 and 2015, http://blog.euromonitor.com/2014/06/the-patterns-of-world-trade.html, © 2014 Euromonitor Virgilijus Narusevicius and Euromonitor International; Figure on page 80 from Accumulated loss of Amazon rainforest 1978–2014, Mongabay.com / Hansen et al 2013–2014, Mongabay.com; Figure on page 81 from Located divided bar graphs to show the origins of water supply in the UK http://www. acegeography.com/wwd---fresh-water-in-the-uk.html, BGS © NERC, 1998, statistics for the use of groundwater across the UK, Permit Number CP16/074 British Geological Survey © NERC 2016. All rights reserved. UK Groundwater Forum.

Maps
Ordnance Survey Maps on pages 4, 9, 10, 18, 20, 27, 98, 101, 102, 117, 119, 120, 123 © Crown copyright 2016, OS 100030901 and supplied by courtesy of Maps International, created by Lovell Johns Limited, Maps International is a trading name of Lovell Johns Ltd.